Mechanics and Adaptronics

Ziel der Buchreihe ist die Publikation neuer Ansätze in der Mechanik und Adaptronik. Diese Ansätze umfassen zum einen experimentelle Themen als auch solche, die numerischen Charakters sind und erstrecken sich über diverse Größenskalen. Experimentelle, zukunftsweisende Themen beinhalten Untersuchungen auf Mikro- bis hin zur Makroebene und berücksichtigen unterschiedlichste Felder, von der Mechanik über Temperatur bis hin zu unterschiedlichen chemischen Feldern. Auch auf Seiten der numerischen Simulation werden auf Grundlage der experimentellen Untersuchungen neue multiskalige Mehrfeldmodelle entwickelt, kalibriert und validiert. Eine Besonderheit stellt hier die Thematik der Adaptronik dar, die als Ziel die Integration neuer Funktionen in tragende Bauteile durch die Kombination konventioneller Konstruktionswerkstoffe mit aktiven Werkstoffsystemen verfolgt. So werden Funktionswerkstoffe betrachtet, die Energiewandler sind und die die klassischen lasttragenden und formgebenden Funktionen um sensorische und aktorische Funktionen erweitern. Aktorische und sensorische Eigenschaften ermöglichen in Kombination mit adaptiven, schnellen Reglern eine optimale Selbstanpassung an die jeweilige Einsatzumgebung. Durch die Zusammenführung von numerischem und experimentellem Know-how auf mechanischer als auch adaptronischer Seite ist es möglich, Erkenntnisse aus der Natur im Sinne des bionischen Gedankens in die Technik zu übertragen und so neue wissenschaftliche Richtungen aufzuschlagen.

Sarah Bornemann

Development of a Self-Sufficient, Wireless Sensor Node Using the Outer FML Layer as an Antenna

Springer

Sarah Bornemann
Gnarrenburg, Niedersachsen, Germany

Veröffentlichung als Dissertation
Promotionsort Universität Bremen
Bremen, Germany

ISSN 2731-6211 ISSN 2731-622X (electronic)
Mechanics and Adaptronics
ISBN 978-3-031-70647-9 ISBN 978-3-031-70648-6 (eBook)
https://doi.org/10.1007/978-3-031-70648-6

This Springer imprint is published by the registered company Springer Nature Switzerland AG
The registered company address is: Gewerbestrasse 11, 6330 Cham, Switzerland

Universität Bremen

Fachbereich 1: Physik und Elektrotechnik
Institut für Mikrosensoren, -aktoren und -systeme

Zur Erlangung des akademischen Grades eines
Doktor-Ingenieur (Dr.-Ing.)
Genehmigte Dissertation

Development of a self-sufficient, wireless sensor node using the outer FML layer as an antenna

von
Sarah Bornemann
aus Gnarrenburg

Referent: Dr.-Ing. Walter Lang
Korreferent: Dr.-Ing. Michael Sinapius

Eingereicht am: 20. Juli 2023
Datum des Promotionskolloquiums: 25. Oktober 2023

IMSAS
Institut für Mikrosensoren, -aktoren und -systeme

Folgende Änderungen weichen von der Dissertation im Original ab:
List of changes
- Figure 1.1 modified
- Equation 4.5 and 4.6 revised
- References listed chapter-wise
- Figures and tables shifted to top/bottom of page, capture position adjusted accordingly to publisher format
- Abbreviations: minutes -> min, Figure -> Fig., Section -> Sect., Chapter -> Chap.

Acknowledgements

In memory of my beloved father.

At this point I would like to thank all the people without whom this work would not have been possible:

First of all I would like to express my gratitude to Prof. Walter Lang for giving me the opportunity to join the Institute for Microsensors, -actuators and -systems (IMSAS) to work on this interesting research topic and for assisting me with valuable discussions while at the same time providing sufficient freedom to develop my own ideas. I would also like to thank Prof. Michael Sinapius for initiating and leading the research unit FOR3022, funded by the German Research Foundation (DFG), being the framework of the work presented in this thesis and also for being my second reviewer.

I am also grateful to Prof. Björn Lüssem for his efforts toward a continuation of the second funding period and the provided opportunity to continue research in this direction. Many thanks to him and Prof. Steffen Paul for being part of the examination committee for the colloquium.

Additional thanks go to all former and present colleagues from IMSAS and the research unit for motivation, inspiring discussions, countless coffee breaks, and unforgettable years in good company! I would also like to thank all the students who did their project and thesis work under my guidance or worked for me as assistants for their invaluable contribution to my work.

Last but not least I want to thank my parents for always motivating me to do anything that came to my mind and especially thank my husband, Sascha Bornemann, for his patience and support in all aspects of my life.

Contents

Abbreviations

AC	Alternating Current
ADC	Analog-to-Digital Converter
ASIC	Application-Specific Integrated Circuit
AWG	Arbitrary Waveform Generator
CFRP	Carbon Fiber-Reinforced Polymer
DAC	Digital-to-Analog Converter
DC	Direct Current
DMA	Direct Memory Access
DUT	Device Under Test
EEPROM	Electrically Erasable Programmable Read-Only Memory
FEM	Finite Element Method
FML	Fiber Metal Laminate
FRP	Fiber-Reinforced Polymer
GBWP	Gain-Bandwidth Product
GFRP	Glass Fiber-Reinforced Polymer
GLARE	GLAss REinforced laminate
GPIO	General Purpose Input Output
GUW	Guided Ultrasonic Wave
HF	High Frequency
IC	Integrated Circuit
LDO	Low Dropout Regulator
MSB	Most Significant Bit
NDT	Non-Destructive Testing
PCB	Printed Circuit Board
PLA	PolyLactic Acid
PTFE	PolyTetraFluorEthylen
PWAS	Piezoelectric Wafer Active Sensors
PWM	Pulse-Width Modulation
PZT	Lead Zirconate Titanate
RAM	Random-Access Memory
RF	Radio-Frequency

RFID	Radio-Frequency IDentification
SAR	Successive Approximation Register
SHM	Structural Health Monitoring
UV	Ultra Violet

Chapter 1
Introduction and Overview

The interest in using Fiber-Reinforced Polymers (FRPs) and Fiber Metal Laminates (FMLs), a combination of FRP with thin metal sheets, has increased significantly in numerous domains over the last decades [1]. Especially for aircraft manufacturers, these materials are a topic of great importance given their advantages compared to the classically used aircraft material aluminum. For the Airbus A380, a large number of parts like wings, tail part, fuselage, and more are made of FML (GLAss REinforced laminate (GLARE)) or FRPs (see Fig. 1.1). The advantages provided by these materials are for example the weight-saving potential, improved fatigue behavior and damage tolerances, the inherent resistance to corrosion, and good fire resistance [2–4].

Thereby, FML even eliminates some known complications of working with pure FRP. While pure FRP is quite brittle by nature, FML can handle impact damage much better because some of the energy leads to plastic deformations in the metal content instead of shattering the fiber layers [4]. At the same time cracks starting in the metal layers can be stopped or bridged by fiber layers, preventing possible structural failure [3]. Therefore, both FRP and FML are very promising materials for components having to fulfill high performance requirements in aircraft applications [5, 6]. The main drawback besides higher material prices is, however, that some damage types can occur during service times and even during the manufacturing process which are not or only barely visible by optical inspection. During the service hours of an airplane, e.g., bird strike, hail, or runway debris, which hit the airplane as a low-velocity impact, can result in hidden cracks and delaminations inside the laminate stack.

Having a plate-like structure lying inside a laboratory, those types of damage can easily be detected using multiple methods, e.g., X-Ray imaging or air-coupled ultrasonic scans. However, using these techniques on a complete airplane during one of the lesser inspection intervals where the parts are not completely accessible from both sides can prove difficult, not to mention the voluminous measurement

S. Bornemann, *Development of a Self-Sufficient, Wireless Sensor Node Using the Outer FML Layer as an Antenna*, Mechanics and Adaptronics,
https://doi.org/10.1007/978-3-031-70648-6_1

Fig. 1.1 Airbus A380 material composition (own representation according to [5])

setup needed for this. In addition to this, on most of the load-bearing structures, reinforcing elements, i.e., stringers are present, preventing or at least complicating the look inside the underlying material.

Facing these challenges, a material-integrated system that could be used to monitor the structure it is embedded in, without the need to disassemble the airplane, would be well suited. Combined with a wireless approach, this is expected to save costs for maintenance without increasing the already significant amount of wire weight for sensor connections in modern aircraft and eliminating the need for connectors that are prone to corrosion. Luckily, there is a possibility to realize embedded Structural Health Monitoring (SHM) using Lamb waves that have already been investigated for such applications in the past. Lamb waves are a type of Guided Ultrasonic Waves (GUWs), which can travel over comparably large plate-like structures. If this wave hits a difference in acoustic material impedance, i.e., a delamination or crack, it changes its characteristics, which can then be measured and used for the detection of such damage. Lamb waves can be introduced into the plate using piezo-ceramic actuators using the piezoelectric effect to convert a time-varying voltage difference into mechanical movement. They offer the advantage that they can also be used as sensors, using the piezoelectric principle the other way around. As an alternative, just recently, a new sensor type was developed based on a completely different working principle and proven to be able to pick up Lamb waves inside the laminate as well [7]. This sensor was developed by another working group within the research unit, under which also this work was created. This sensor provides the advantage that it is completely lead-free and so small that it can be easily integrated into the material. A more detailed description of Lamb waves, their behavior, and the sensing approaches will be provided in a later chapter.

The focus of this work will be the development of a wireless electronic sensor node that can be used for the acquisition of Lamb wave sensor data and can be fully embedded into the structure. The sensor node will be able to use either one of the given sensor principles to measure Lamb waves inside an FML plate. This is realized by choosing a fully digital approach. Despite the digital processing of the sensor signal, the components are chosen to be as thin as possible to create a sensor node that disturbs the laminate by the least possible amount. The wireless interface will be realized by structuring an antenna into the outer layer of the FML and compensate for the effects caused by the partly conductive environment the antenna is expected to work in.

One of the main issues will be to investigate how much electronic function can be realized in a passive, material-integrated system without any permanent power source but a temporary available energy, harvested from the reader field during sensor readout. Therefore, performance measurements will be conducted, setting the implemented functions, the speed of execution, and the overall power consumption of the system in relation. One major aim is to be able to assess the margin for further operational scenarios of the developed system with regard to a possible implementation of SHM damage detection algorithms or similar advanced software on the system.

Material integration will be done given challenging manufacturing conditions exceeding the specified limits of electronic components. The resilience to these conditions will be tested extensively beforehand. Furthermore, a wireless connection approach will be chosen in a way that a power supply can be realized by harvesting the wireless energy present anyway during the read-out of the sensor node. The antenna for the wireless connection will be designed in a way that it can be later on structured into the outer metal layer of an FML component using a suited structuring approach. The additional challenge in the design of the wireless connection resulting from this is, that it will be required to not only work on glass fibers but also on carbon fibers, which due to their conductivity influence the antenna's performance. The impact of conductive fibers, their orientation as well as approaches to face this challenge will be investigated within this work as well.

Finally, the complete system functionality will be demonstrated by material-integrated measurements of Lamb waves inside FML plates. Measurements will be done in undamaged plates and with applied pseudo defects to be able to detect changes in the measured signal responses at the sensor location as a first step toward material-integrated damage diagnostics. Regarding reproducibility of the results, for pseudo-damage, it will also be shown, that after the damage is removed, the measurement shows the baseline of an undamaged plate again.

1.1 System Overview

This Section is intended to first give a short overview of the components of the wireless sensor node and list the requirements that have to be considered during its design process. A short overview of the system components is provided, which are later on explained in more detail. References to the listed requirements can be found in Chapter 4 in particular, as it presents the design process and the decisions made therein.

1.1.1 Components of the System

The presented system consists of an antenna, a small electronic sensor node, a sensor connected to it, and the laminate itself into which the aforementioned components are embedded (see Fig. 1.2). To complete the system, piezo ceramic actuators are used to excite Lamb waves inside the material and a reader unit connected to a computer allows for data processing, plotting, and assessment. The last two are not developed within this work and will just be introduced once when the measurements performed with the wireless sensor node are presented.

1.1.2 Requirements Analysis

A requirements analysis is of great importance whenever a system is about to be designed. It acts as a base for discussion between system designers and system users or multiple designers in the first place but also reveals optimization challenges whenever there are two opposing design constraints present. The system within the scope of this work is not highly complex, but still, some requirements have to be kept in mind during the design process (see Fig. 1.3).

The material integration requirement is the main reason for many design decisions made within this project in the first place. It is influencing the electronics as well as the antenna in a great amount. Leaving the fiber type open for both, non-conductive glass fiber as well as conductive carbon fiber reinforcements, the electronic sensor node has to be designed in a way that it is not electrically shorted by this. The antenna at the same time has to be modified to work on conductive surfaces as well. Assuming that the conductivity changes accordingly to the layer setup and the fiber volume content, an insulation method has to be identified, that deals with this. At the same time, electronics and antenna have to be chosen as material compliant as possible by choosing a suited geometry on one hand and keeping dimensions in all directions small. Lastly, the material integration also generates new requirements regarding the embedding conditions under which the material is fabricated. Following the fabrication process steps, provided by the manufacturer, is vital to ensure material performance. This means that all components that are brought into the laminate stack

Fig. 1.2 Schematic representation of the system idea showing the antenna inside the outer FML layer and the sensor node embedded inside the material stack

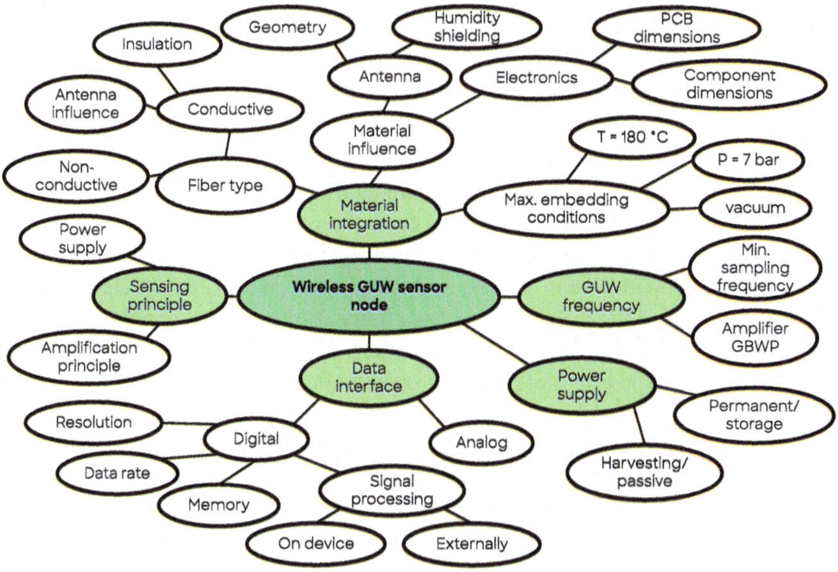

Fig. 1.3 Requirements overview for a wireless GUW sensor node

before component manufacturing have to be tested to survive this process at elevated pressures up to 7 bar, combined with temperatures of up to 180 °C, well above their specified storing conditions.

Another design constraint that mainly matters for the electronics part is the frequencies of the Lamb waves that will be used for SHM later on. This determines the minimum sampling frequency of the system and limits the amplifying components that come into consideration. This also influences the decision of whether an analog or digital approach for the system is better suited. Further, a decision has to be made concerning the power supply. Is an energy storage required to create a permanently active system or will a passive system, that only wakes up when it is required, be sufficient?

The last two main requirements are the sensing principle chosen, which influences the need for a provided voltage, as needed in case of, e.g., a bridge measurement, and also the required amplification principle. Taking piezo ceramic sensors as an example, a charge amplifier would be well suited, while for other approaches this is not suited. Finally, the applied sensing principle itself also influences the method for signal processing, analog or digital. This then further influences the possibilities for the use of the sensor node in a later application. Designing a fully digital approach offers new possibilities, for example enabling rudimentary damage prediction, implemented on the sensor node, at the location of events.

1.2 Structure of This Work

This work is structured into six chapters. The three main chapters follow the same scheme. They are divided into a presentation of the current state-of-the-art for that specific chapter at the beginning, followed by the theoretical background. Finally, each chapter presents the main content, containing design processes, simulations, experimental setups, measurements, and evaluations of the results.

In the second chapter a closer look is taken at the material of interest, FML, into which the developed system will be embedded later on. It is considered to give a general overview of the material composition, its characteristics, and manufacturing methods and to introduce some of the challenges faced during the system design process. Also, the materials and material combinations used within the frame of this work will be introduced here.

The third chapter addresses the antenna part of the system including chosen working frequency, the design process based on theoretical calculations, and its iterations based on measurements. Mechanical considerations that are influencing the design process are presented here as well. This chapter also introduces possible solutions to the challenge of conductive fibers coming from one of the requirements listed in Sect. 1.1.2 and explains why the chosen solution is the best-suited one for this specific application scenario. Antennas are manufactured and the actual influence of the application environment is measured for different sizes, shapes, and underlying materials. After this, the energy harvesting capabilities of different antenna sizes are evaluated and thereby directly narrowing the design space for power dissipation for the electronic system to be developed. Finally, to be able to investigate the influence of different materials around the antenna and its performance before a new design is manufactured, a finite element method simulation is set up using COMSOL multiphysics.

Chapter four presents the design process for the sensor node itself. For every component the reasons why exactly this component is the best-suited one will be provided. Considerations on the material integration process, as well as limits for material integration of commercial electronic components, will be presented. Two different software implementations, enabling the sensor node to digitize sensor data and provide the raw values for readout are introduced. One of them provides the additional possibility to use a software trigger, thereby offering the advantages of averaged measurements, is presented. Characterization of the final sensor node regarding the resolution of the signal and current consumption depending on the functions implemented and the achievable sampling rate will be presented here as well.

Finally, in chapter five, SHM focusing on Lamb wave measurements is introduced and material-integrated measurements are performed using the developed wireless sensor node. The readout side of the system and the data processing after readout are shortly introduced here as well. The focus of this chapter will be to perform material-embedded measurements and to investigate if it is possible to distinguish the baseline signal from a damage signal by capturing the small significant changes wirelessly measured with both sensors alike. The first concern is the quality of measured signals.

Therefore, the sampling rate will be evaluated for multiple Lamb wave excitation frequencies. The second question, if reproducible measurements can be performed, is evaluated by applying magnets, representing easily removable pseudo defects, and removing them multiple times, measuring Lamb wave propagation.

The last chapter sums up the results of this work and provides an outlook on the next steps that can be made toward the creation of a completely material-integrated, wireless SHM system.

References

1. M. Mrazova, Advanced composite materials of the future in aerospace industry. Incas Bull. **5**(3), 139–150 (2013)
2. A. Asundi, A.Y.N. Choi, Fiber metal laminates: an advanced material for future aircraft. J. Mater. Process. Technol. **63**(1), 384–394 (1997)
3. G.B. Chai, P. Manikandan, Low velocity impact response of fibre-metal laminates - a review. Compos. Struct. **107**, 363–381 (2014)
4. L.B. Vogelesang, A. Vlot, Development of fibre metal laminates for advanced aerospace structures. J. Mater. Process. Technol. **103**(1), 1–5 (2000)
5. S. Kesarwani, Polymer composites in aviation sector - a brief review article. Int. J. Eng. Res. Technol. **6**(06) (2017)
6. M. Chandrasekar, Ishak, M. Jawaid, Z. Leman, S.M. Sapuan, An experimental review on the mechanical properties and hygrothermal behaviour of fibre metal laminates. J. Reinforced Plastics Compos. **36**(1), 72–82 (2017)
7. J.N. Haus, W. Lang, T. Roloff, L. Rittmeier, S. Bornemann, M. Sinapius, A. Dietzel, Mems vibrometer for structural health monitoring using guided ultrasonic wave. Sensors **22**(14) (2022)

Chapter 2
Fiber Metal Laminates

FMLs combine thin layers of metal sheets alternating with layers of FRPs in an adhesive bond. Depending on their field of use, they can be adjusted to the application-specific requirements. In aerospace applications, for example, they provide increased resistance to impact damage, corrosion, and metal fatigue compared to commonly used aluminum [1].

By varying the fiber directions, they can either be designed for high specific strength in one direction or distributed strength in multiple directions. At the same time, the metal layers protect the underlying FRP from environmental influences like humidity and impact damage while reducing the lightweight-construction potential of pure FRP just slightly. The main drawback of using FML instead of aluminum is the higher initial manufacturing cost but for many applications this can be relativized by longer lifetime estimations and the advantage of increased freedom at part design and thereby fewer parts.

For the metal layers, almost any type of metal can be used. Especially aluminum and stainless steel have been used for research and industrial applications lately [2–5]. For the FRP part even a bigger variety of material combinations is available. Chandrasekar et al. report that the combination of three metal layers and two fiber composite layers (3/1), as well as the (2/1) combination with two metal and one FRP layer, are the most common ones in research publications [2]. The configurations and manufacturing techniques for FML components as well as the other materials used within the frame of this work will be introduced in the following sections.

2.1 Fiber Reinforced Polymers

FRPs are a hybrid material that consists of reinforcing fibers surrounded by a resin matrix. Both parts are adhesively bonded during the manufacturing process. For load-bearing FRP structures, usually continuous fibers are chosen as reinforcements.

S. Bornemann, *Development of a Self-Sufficient, Wireless Sensor Node Using the Outer FML Layer as an Antenna*, Mechanics and Adaptronics,
https://doi.org/10.1007/978-3-031-70648-6_2

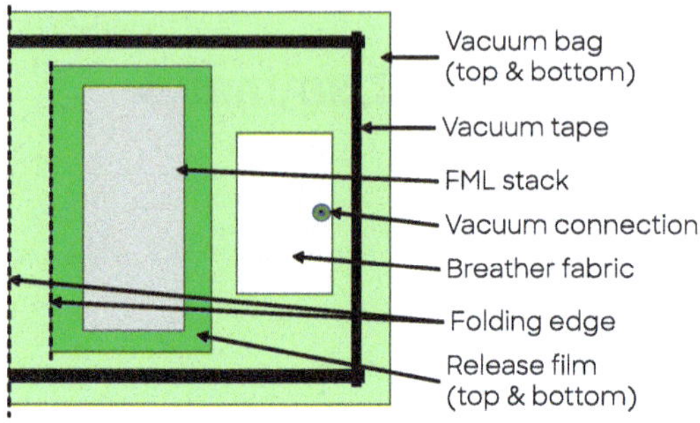

Fig. 2.1 Fiber orientations during stacking sequence of FRP and FML manufacturing

Glass fibers, aramid fibers, carbon fibers, and hybrids from these materials are most commonly used for aerospace applications nowadays [6].

Fiber composite parts can be manufactured by a variety of methods, e.g., pultrusion, wet layup, vacuum infusion, and autoclave manufacturing, depending on the available machinery, the final application, and the chosen material. In the aerospace industry, autoclave manufacturing is the most often used method. For this procedure, pre-impregnated layers of fibers, also known as prepregs, are used as the basic material. The layers of the FML-stack are cut, stacked, and covered with a release film on both sides. A breather fabric can either be wrapped around this stack or put close by to enable the air to be removed from the vacuum bag completely (see Fig. 2.1). The bag is closed using vacuum sealing tape and the complete setup is placed into an autoclave. During manufacturing, the recommended curing cycle for the selected prepregs is automatically run by the autoclave. During that time it precisely controls vacuum level, pressure, and temperature.

The same principle can be used in a heat press. Here, the prepared vacuum bag is placed between heated press jaws which apply the predefined amount of pressure to the material stack mechanically while heat-curing the laminate in between. Depending on the thickness of the material stack, both sides have to be heated. For small stacks up to 2 mm heating from one side has experimentally proven to be sufficient with an increase of manufacturing time of 10 min.

2.2 Fiber Orientation and Material Influence

The fiber orientation within the FRP layers of the FML has a significant impact on mechanical strength as well as the electrical properties of the finished component. It is usually chosen depending on the application and mechanical requirements of the component and can vary over the different parts of a final product, e.g., an airplane. Orientation definitions are used at several locations in this work. To maintain consistency, Fig. 2.2 shows the generally used definition of stacking sequences for FRP.

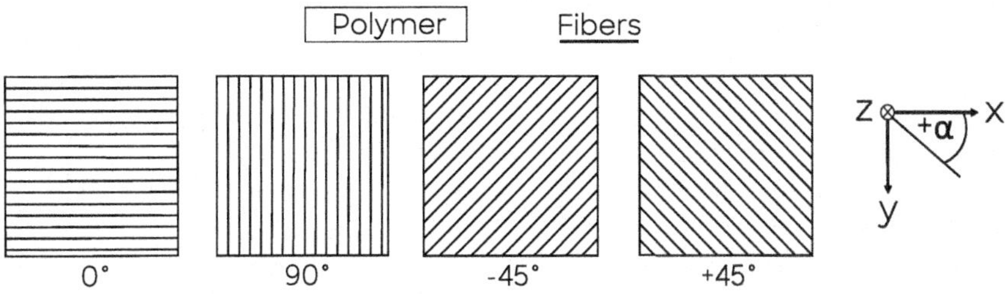

Fig. 2.2 Fiber orientations of FRPs used within this work

Due to high stiffness of the fibers unidirectional FRPs show very high strength when forces occur in parallel to the fiber direction, i.e., in 0° direction, because the matrix material helps to evenly distribute the force among multiple fibers at the same time. For perpendicular forces, the strength of the compound material is significantly smaller, somewhere in the range of the pure matrix material [1].

For carbon fiber reinforced components, the stacking sequence not only influences the mechanical but the electrical properties of the FML as well. The unidirectional stack-up usually provides significantly higher electrical conductivity in fiber direction than in the transverse directions, making it a very anisotropic conductor in total. The quasi-isotropic setup on the other hand, using a combination of 0°, 90°, 45° and −45° layers alike, results in isotropic properties in x-y direction, both mechanically and electrically. The stacking sequences used for the experiments in this work can be found in Table 2.1.

Table 2.1 Pure CFRP plates used for antenna performance experiments. Quadratic dimensions w=100 mm, material *HexPly® 8552*

Layer no.	Unidirectional	Perpendicular	Multidirectional
1	0°	0°	45°
2	0°	90°	−45°
3	0°	0°	45°
4	0°	90°	−45°
5	0°	0°	0°
6	0°	90°	0°
7	0°	90°	0°
8	0°	0°	0°
9	0°	90°	0°
10	0°	0°	−45°
11	0°	90°	45°
12	0°	0°	−45°
13	-	-	45°

2.3 Materials Used in This Work

For the plate manufacturing for experimental results presented within this work, three different types of prepregs and two types of metal sheets are used. One Carbon Fiber-Reinforced Polymer (CFRP) prepreg is the high pressure, high-temperature curing, *HexPly® 8552*, recommended for the construction of primary structure components in airplanes. The prepreg consists of an epoxy matrix, the fiber type is carbon fiber *HexTow® AS4* with a fiber diameter of 7.1 µm, carbon content of 94.0 % and an electrical resistance of 1.7 mΩ cm [7]. The CFRP part of the cured FML using the *HexPly® 8552* will contain 57.42 % fiber volume content according to manufacturer information. This pre-preg is cured using an autoclave at the Institute of Mechanics and Adaptronics (ima) at the TU Braunschweig using the manufacturer-recommended curing cycle, step-wise listed below, taken from [8].

1. Full vacuum (1 bar)
2. 7 bar gauge pressure
3. Reduce vacuum to 0.2 bar when pressure 1 bar gauge
4. Heat at 1–3 °C min^{-1} to 110 °C ± 5 °C
5. Hold for 60 ± 5 min
6. Heat at 1–3 °C min^{-1} to 180 °C ± 5 °C
7. Hold for 120 ± 5 min
8. Cool at 2–5 °C min^{-1}
9. Vent pressure when temperature ≤ 60 °C

The second type of prepreg is Glass Fiber-Reinforced Polymer (GFRP), representing a non-conductive alternative cured at a lower temperature of only 120 °C, that is used for the first embedding experiments. It represents the currently mostly used FML combination GLARE, combining GFRP layers and thin aluminum sheets. The plates are manufactured in a double-sided heat press at the Faserinstitut Bremen e.V. (FIBRE).

The last type of FRP prepreg used within this project is the *UNIPREG® Kohlegelege* 150g/m^2. It is another type of CFRP, meant to be used for highly loaded structural parts but is cured at 120 °C [9]. It is mainly used for the experiments on climate influence of antennas.

For the metal part, two types of metal are used. In combination with GFRP, 400 µm thick aluminum sheets are used, while for all experiments on CFRP combinations, 100 µm thick austenitic stainless steel foil of type *1.4310 / X10CrNi18-8* is used. Stainless steel represents a challenging material from an antenna point of view, given its low conductivity compared to other antenna conductors. Whereas aluminum combined with layers of GFRP is used to investigate the suitability of the developed system also for the most popular FML type, GLARE, already used in aviation nowadays, e.g., in the Airbus A380.

References

1. F. Campbell, *Structural Composite Materials* (A S M International, Materials Park, 2010)
2. M. Chandrasekar, Ishak, M. Jawaid, Z. Leman, S.M. Sapuan, An experimental review on the mechanical properties and hygrothermal behaviour of fibre metal laminates. J. Reinforced Plastics Compos. **36**(1), 72–82 (2017)
3. E. Hariharan, R. Santhanakrishnan, Experimental analysis of fiber metal laminate with aluminium alloy for aircraft structures. Int. J. Eng. Sci. Res. Technol. (2016)
4. Q. Liu, J. Ma, L. Kang, G. Sun, Q. Li, An experimental study on fatigue characteristics of cfrp-steel hybrid laminates. Mater. Des. **88**, 643–650 (2015)
5. T. Pärnänen, M. Kanerva, E. Sarlin, O. Saarela, Debonding and impact damage in stainless steel fibre metal laminates prior to metal fracture. Compos. Struct. **119**, 777–786 (2015)
6. Hexcel Corporation, *Aerospace Selector Guide* (2022). Product Selector Guide
7. Hexcel Corporation, *HexTow AS4* (2020). Datasheet
8. Hexcel Corporation, *Hexply 8552* (2020). Datasheet
9. R&G Faserverbundwerkstoffe, *Technische Daten UNIPREG Kohlegelege 150 g/m*^{2} (2022)

Chapter 3
Wireless Interface

Antennas are used in nearly all parts of today's daily life. Entertainment programs are received from radio stations, app updates using WIFI or 5G on smart devices, or the Bluetooth connection to our headphones. Not only data can be transferred without wires, power can be transmitted as well which is increasingly used in the entertainment industry recently, e.g., wireless charging stations for smartphones and smartwatches. The antennas used differ significantly in size and shape and even on their physical working principle depending on their application.

This chapter will provide a short overview of the state of the art focusing on applications where sensor nodes embedded into materials are connected to the outside world using a wireless link, discriminating mainly two power supply approaches, physical energy harvesting and using the energy from the data link as a supply. Different possibilities for simultaneous energy and data transfer options for wireless sensor applications are introduced, compared, and the important equations for the chosen approach are presented within the theoretical background section. After that the iterative design process including details on manufacturing, contacting, mechanical simulations, and design calculations is presented, taking the challenging application area close to multiple conducting materials into account. Finally, a simulation model is set up to allow for a faster decision on design adjustments due to changing requirements on one side and to easily simulate other influences, e.g., a different conductor material on antenna performance without experimentally measuring every configuration. The main results from this chapter have been published in [1].

S. Bornemann, *Development of a Self-Sufficient, Wireless Sensor Node Using the Outer FML Layer as an Antenna*, Mechanics and Adaptronics,
https://doi.org/10.1007/978-3-031-70648-6_3

3.1 State-of-the-Art: Wireless Sensor Nodes

The integration of sensors into multiple materials has also increased the research toward wireless approaches for these systems. For the data transmission side, this can be implemented by simply using one of the already existing and optimized technologies like wireless local area network, Zigbee, Bluetooth, and many others. Building on these communication techniques, powerful, long-range sensor networks can be designed. Combining one of the aforementioned communication approaches with a permanent energy supply which requires wires toward the sensor system defeats the purpose of setting up a wireless system. Therefore, in this work, a fully wireless approach for data and power is investigated.

3.1.1 Wireless Sensor Node Supplies

One possibility is the use of an energy storage, e.g., a battery. This comes with the necessity to replace or recharge the battery every now and then during the system's lifetime. Besides that, relying on energy buffers for system supply contains the danger of being a major lifetime limitation. But for systems that are still somehow accessible after material integration, this may be a suited option nevertheless.

Energy harvesting is a very attractive alternative for wireless sensor node supply and in many applications, there are sources from which energy can be potentially harvested. For example, in areas of daily life, a small amount of energy can be gained from the anyway present wireless networks. Dhar et al. show that more than 600 µW of real-time power can be harvested from a 2.4 GHz router in 30 cm distance [2]. While Alneyadi et al. show in [3], that even in a distance of 104 cm, a power of about 1.26 µW can be extracted by using two combined patch antennas and a frequency of 2.36 GHz. Microwatts of power are however not sufficient to power most sensor nodes directly, but for these approaches harvesting some amount of energy over time and storing it works, to supply a low-power system for a few seconds.

A different approach can be found in wireless sensor systems which rely completely on energy harvesting during their operational time. They harvest energy from the communication field, without storing it for later usage. This also means that these systems are only active during operation and remain completely passive when no energy harvesting source is available. This comes with the clear advantage that no energy storage has to be implemented but this is also accompanied by some disadvantages. Harvesting energy decreases the operating range of such systems significantly and, in addition, the sensor data processing and possible storage have to be completely finished before the energy harvesting source is removed again. Given Radio-Frequency IDentification (RFID) as an example, this can be achieved by making the device that provides the energy, i.e., the magnetic field also giving out the data. With this, the user is forced to wait until the sensor value has been read and transmitted before removing the reader device again. Several systems can be found which are relying on this method as the best one suited.

Jiang et al. [4] describe an analytical method to compute the power delivery as a function of mutual inductance and provide guidelines for the design procedure of a power harvesting circuit. Gould et al. [5] use an inductive power transfer at 13.56 MHz to wirelessly power a temperature sensor for condensation detection which requires a maximum of 2 mW of power for operation. In a survey of NFC sensors based on energy harvesting, High Frequency (HF)-RFID is even specified to deliver approximately 10 mW of power, in contrast to Ultra-High-Frequency RFID with only a few μW [6]. The review paper by Costa et al. [7] presents the latest progress in RFID sensors for all RFID operating frequencies and provides a good overview of the possibilities these systems offer for Internet of Things applications.

3.1.2 Antennas on Conductive Materials

A big challenge in using wireless links for sensor nodes is the fact that in contrast to wired connections, wireless approaches are often much more vulnerable to environmental influences and changes. Several publications can be found for investigations around on-metal RFID tags, focusing mainly on HF frequency applications.

Bovelli et al. [8] describe the detuning influence of metal underneath the tag antenna in dependence on the distance between metal and antenna in their work, proposing the antenna coil wound around a ferrite core and fixing it on a known metallic sheet, which can then be placed on other metallic surfaces as a solution for a functioning antenna on metal. Qing et al. [9] as well as Petrariu et al. [10] investigate the influence of metal sheets of different sizes on metal backed loop antennas, working in this constellation with a combination of stable spacing to the metal and retuning to the resonance frequency with the help of additional passive components. The approach of stable distance to the metallic surface is also taken in the mathematical investigation of on-metal applications of HF-RFID antennas in [11].

Carbon fibers in CFRP result in similar challenges as metal layers because of their conductivity. Nevertheless, some works investigate the possibility to attach metal-backed antennas onto CFRP or even embed them into FRP with layers of carbon-reinforced composite materials surrounding them. Mayordomo et al. [12] investigate the influence of conductive fibers inside CFRP on the performance of attached HF-RFID antennas, with a comparable result as seen for the application on metal surfaces. The antenna resonance frequency is shifted to a higher value and the antenna does not work in close proximity to CFRP material. They provide a simulation of the given correlations and a solution for this again keeping a distance to the conductive material of 5 mm combined with retuning.

A different approach can be found in the works of [13–15]. All of them describe the detuning and attenuating effect of CFRP beneath antennas and apply a high permeability material, i.e., a ferrite sheet between the conductive surface and antenna to counter this effect. Salas et al. even show that for low-frequency transmission at 125 kHz a transmission to an antenna inside a pocket of CFRP is possible with just a slightly decreased coupling factor compared to an application in air. The combination of a ferrite sheet with a metal backplane to create a constant influence of the antenna for a later application on various surfaces is presented by Gebhart et al. in [16].

An even more challenging material for the use of material-integrated antennas is represented by FML. As already introduced in Sect. 2, in addition to fiber composite layers, which might already contain conductive carbon fibers as reinforcements, there are alternating metal layers in it. The top and bottom are thin metal layers for this hybrid material. Therefore, the integration of an antenna inside the material is not possible. The idea of this project to structure the antenna coil into the outer metal layer could not be found in literature, besides our own publications, so far.

3.2 Theoretical Background: Wireless Interface

In this section, the theoretical background necessary for the antenna design process and carried out optimization steps is presented. Therefore, it reaches from general antenna working principles over different communication standards and calculations for the antenna design toward an introduction of different available measurement methods, gaining importance for the upcoming chapters evaluating the antenna's performance. An additional section deals with the water absorption behavior of materials. This will be of interest for the experimental investigation of climate influences on the underlying material, exposed by structuring the antenna into the top layer.

3.2.1 *Maxwell's Equations*

Maxwell's equations describe the cause and interrelation of electric and magnetic fields. All wireless interfaces work on the basis of these theoretical principles. The four of Maxwell's equations in matter are represented in differential form in the following:

The most important one for this work

$$\nabla \times \boldsymbol{E} = -\frac{\partial \boldsymbol{B}}{\partial t} \tag{3.1}$$

is Faraday's law of induction, which states that changing magnetic flux densities $\boldsymbol{B}$ are the cause of electric fields $\boldsymbol{E}$. The second most prominent equation is an extension of Ampères law

$$\nabla \times \boldsymbol{H} = \boldsymbol{J} + \frac{\partial \boldsymbol{D}}{\partial t} \tag{3.2}$$

and states that electric current density $\boldsymbol{J}$ as well as changing electric displacement fields $\boldsymbol{D}$ are the cause for magnetic fields $\boldsymbol{H}$. The third one

$$\nabla \cdot \boldsymbol{B} = 0 \tag{3.3}$$

states that the total magnetic flux density through a closed surface has to be zero, thereby no magnetic monopoles exist and magnetic flux lines always form closed loops. The fourth equation

$$\nabla \cdot \boldsymbol{D} = \rho_c \tag{3.4}$$

states that the electric displacement field over a closed surface is equal to the enclosed volume charge density ρ_c [17, 18].

The vectors of the magnetic field and the displacement field can be found because these equations are the so-called macroscopic equations that take into account that the application does not happen in a vacuum but the field lines will pass through other materials as well. With this, the relation between the magnetic field and magnetic flux density is given by $\boldsymbol{B} = \mu \boldsymbol{H}$, whereas the electric flux density is related to the electric field by $\boldsymbol{D} = \varepsilon \boldsymbol{E}$ [17, 18].

Some of the observed correlations in this work can be justified by Maxwell's equations, for example, the effect conductive material has on the antenna functionality.

3.2.2 Antenna Frequency and Free-Space Loss

Antennas are designed accordingly to their targeted frequency of use. For electromagnetic waves, this is associated with the wavelength of the signals transmitted or received. The correlation can be described by Eq. 3.5, where λ represents the wavelength, f is the frequency of the signal, and v_p is the phase velocity of the medium through which the wave propagates. For the case of vacuum, v_p equals the speed of light c.

$$\lambda = \frac{v_p}{f} \tag{3.5}$$

The reasons why RF energy harvesting systems at higher frequencies work at larger distances but are only able to harvest a comparably small amount of power from the RF field can be found in a fast decay of the field strength over distance. The fundamental equation that describes this is Friis transmission formula [19]

$$P_r = P_t \cdot D_t D_r \left(\frac{\lambda}{4\pi R} \right)^2 \tag{3.6}$$

where R is the distance, P_t and P_r, and D_t and D_r are the power and directivity of the transmitting and receiving antenna, respectively. Based on this equation, the ratio of transmitted to received power is defined as the free-space path loss [20] and can be given for two isotropic, non-directive ($D_t = D_r = 1$) radiators as

$$FSPL = \left(\frac{4\pi R}{\lambda}\right)^2. \tag{3.7}$$

Hence, at the same frequency, the transmitted power degrades in dependence of R^2 when increasing the distance between both antennas. This influence increases for higher frequencies, resulting in a smaller wavelength and thereby in an even higher free-space path loss, scaling with λ^2. This means in order to receive the same amount of power from a signal at ten times the frequency, 100 times the input power to the transmitting antenna would be required to achieve this. For real applications, at this point usually official regulations are preventing systems to provide unlimited transmitted power and thereby limiting the power that can be harvested at higher frequencies to a certain amount. This also explains the publications listed in Sect. 3.1.1 for GHz energy harvesting, which are always found to charge an energy buffer over some time to enable a low-power system to work periodically for a few seconds. In addition to that, the distance over which these systems are able to work for energy harvesting is only limited to a fraction of the usual working range of GHz systems.

Therefore, for passive energy harvesting sensor nodes, the choice of working frequency is often a trade-off between data transmission rate and energy harvesting capability at that frequency, in combination with the desired working range in the end. Regarding these correlations, at this point, the different RFID working principles that could be used for the application will be shortly introduced to make a decision on the best-suited one for the targeted application to continue with.

3.2.3 RFID Standards

RFID communication in general is in use since the 1980s for specialized applications, since the 1990s, however, manufacturing improvements and mainly miniaturization successes led to an increased number of possible use cases for such systems [21]. Among other features, the center working frequency can be used to distinguish between the different available systems. Depending on the frequency, also the physical working principle can be different. The quite common distinction into three groups: low frequency, high frequency, and ultra-high frequency will be used here as well, describing each frequency standard in a very condensed form. The idea is to allow for a comparison between the types and to be able to make a justified decision on which approach to use for this project.

Low-Frequency RFID

The low-frequency systems operate between 30 and 300 kHz and use inductive coupling between the reader and transponder coil for energy and data transfer. Due to the low frequency that is used, the influence of liquids and non-conductive objects in the surrounding is negligible. Therefore, those systems are preferred whenever living tissue is placed between the system components, i.e., for animal identification. Lately, low-frequency systems are also increasingly used for power transfer applications, e.g., wireless charging of smartphones, watches, or even cars. The drawbacks of low-frequency RFID systems are however, the comparably low read range of up to 80 cm [22] and a very low data transfer rate of 4–8 kbps [23].

High-Frequency RFID

The high-frequency systems operate at frequencies between 3 and 30 MHz. A majority of devices operate at a center frequency of 13.56 MHz, following either ISO/IEC 14443, i.e., the proximity standard, or the ISO/IEC 15693, i.e., the vicinity standard. With the vicinity standard, in theory, it is possible to reach up to 1 m communication distance for passive devices [22]. This distance is reduced drastically by unfavorable environmental effects as well as energy harvested from the reader field, e.g., for the supply of additionally connected components. The achievable data rate varies between 6.7 and 848 kbps depending on the protocol used, while this approach offers the advantage of multiple tags being read out in the same area due to the implemented anti-collision protocol [23]. An application close to liquids or in tissue is possible, here only a small influence can be noticed. The influence of conductive materials is by far stronger but can be controlled with additional methods which will be discussed later on.

Ultra-High-Frequency RFID

The last category of RFID systems operates at frequencies higher than 300 MHz. The most often used center frequencies are region-dependent 866–868 MHz in the EU and 915 MHz in the US. They offer comparably long reading distances of up to 12 m, small antenna sizes, and higher data rates [22]. At this frequency, electromagnetic coupling in the far-field region around the reader antenna is used, which allows for space-efficient dipole-based antennas but massively suffers from a rapid decay of transferred energy within the far-field region. In addition to that, they are strongly influenced by liquids in the proximity of the antenna. In electrically conductive environments, reflections significantly complicate the use of UHF tags.

3.2.4 Inductively Coupled Antennas

Based on Sect. 3.2.3, as a compromise between data rate, working distance, and energy harvesting capabilities, the HF-RFID standard is chosen for the sensor node developed in this work. Several examples in Sect. 3.1.1 promise a sufficient amount of power to supply a digital sensor node, given a low-power approach. An RFID system consists of four main components, a reader, a transponder, also known as a tag, and both sides' individual antennas. Power and data can be exchanged between the systems wirelessly, as long as the two antennas are close to each other so that the tag antenna is as exposed to the magnetic field induced by the reader antenna as much as possible (see Fig. 3.1).

Due to this working principle, the HF-RFID approach can be better compared to loosely coupled transformer coils than to radiating antennas like dipole or Yagi antennas. Classical antennas produce a radiating electromagnetic field in the far field region, which can travel over long distances, while planar coil antennas for HF-RFID use the magnetic field in the near field around the antenna. Therefore, the equations useful for design and the estimation of transferred voltage are also more like the ones used for a transformer [24].

The working principle of inductive coupling is based on the effect that an electric current passing through a conductor causes a magnetic field in its surrounding. For a single line conductor through which a current I passes, the magnetic field strength around it at a constant distance r can be described by

$$\boldsymbol{H} = \frac{I}{2\pi r} \tag{3.8}$$

with the direction of the magnetic field being clockwise, looking in the current direction. The field strength of a square loop conductor depends on the current through the conductor I, its number of turns N, and its edge lengths a and b and follows the equation [24, 25]

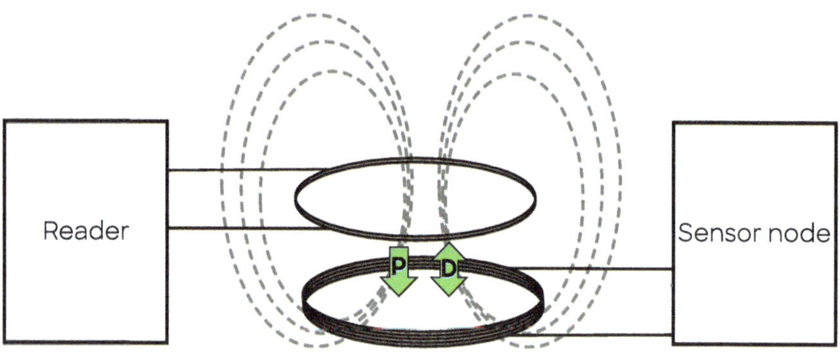

Fig. 3.1 Schematic representation of the RFID working principle showing the field around both, reader and sensor node antenna. Power (P) and data (D) are exchanged wirelessly

$$\boldsymbol{H} = \frac{I \cdot N \cdot ab}{4\pi \cdot \sqrt{(\frac{a}{2})^2 + (\frac{b}{2})^2 + x^2}} \cdot \left(\frac{1}{(\frac{a}{2})^2 + x^2} + \frac{1}{(\frac{b}{2})^2 + x^2} \right) \tag{3.9}$$

where x is the vertical distance from the center. This equation is only valid within the near field around the coil $x < \lambda/2\pi$.

The read range and radius of the antenna are closely coupled. The optimum radius of a transmitting antenna is $\sqrt{2}$-times the maximum targeted read distance, but for too big antennas this can lead to an insufficient field strength very close to the antenna to sufficiently power up tags in this region [24].

The magnetic flux density is a function of magnetic field strength and describes the interaction of an applied magnetic field strength when it passes through a medium. Magnetic field strength and magnetic flux density are therefore coupled to each other by the relative permeability μ_r and the magnetic constant μ_0 of the surrounding material.

$$\boldsymbol{H} = \frac{\boldsymbol{B}}{\mu_r \cdot \mu_0} \tag{3.10}$$

The magnetic flux density that passes through an area A results in the magnetic flux Φ [26]

$$\Phi = \int_A \boldsymbol{B} d\boldsymbol{A}, \tag{3.11}$$

which is directly connected to the third Maxwell's equation. If there is more than one conductor winding N within the same area A, the concatenated flux Ψ results from the sum of all magnetic fluxes, or the amount of windings times a single flux.

$$\Psi = \sum_{i=1}^{N} \Phi_i = N \cdot \Phi = N \cdot \int_A \boldsymbol{B} d\boldsymbol{A}. \tag{3.12}$$

The self-inductance of such a coil can be defined as

$$L = \frac{\Psi}{I} = \frac{N \cdot \Phi}{I}. \tag{3.13}$$

If there are two single loop coils present in close proximity to each other and the magnetic field originating from coil 1 passes through coil 2, the coupling of both results in the mutual inductance [26]

$$M_{21} = \int_{A_2} \frac{\boldsymbol{B}_2(I_1) d\boldsymbol{A}}{I_1} \tag{3.14}$$

which due to the reciprocity theorem is the same for M_{12} and can thereby be called M without any indices.

To express the coupling between two coils without considering their geometries, the coupling factor k between two RFID antennas in distance x to each other and centered on the same axis for $r_{tag} \leq r_{reader}$ is defined by [27]

$$k(x) = \frac{r_{tag}^2 \cdot r_{reader}^2}{\sqrt{r_{tag} \cdot r_{reader}} \cdot \left(\sqrt{x^2 \cdot r_{tag}^2}\right)^3}. \tag{3.15}$$

If one coil is inserted into the magnetic field of another coil, a voltage is generated in that loop that can be calculated by [24]

$$u_2 = \frac{j\omega M \cdot i_1}{1 + \frac{j\omega L_2 + R_2}{R_L}} \tag{3.16}$$

which becomes

$$u_2 = j\omega M \cdot i_1 \tag{3.17}$$

for an open circuit configuration and $u_2 \to 0$ for a small R_L, i.e., a high power requirement of the circuit.

By using resonant inductive coupling instead of regular inductive coupling, by adding a parallel tuning capacity C_{tun} (see Fig. 3.2), this can be improved to

$$u_2 = \frac{j\omega M \cdot i_1}{1 + (j\omega L_2 + R_2) \cdot \left(\frac{1}{R_L} + j\omega C_{ges}\right)} = \frac{j\omega k \cdot \sqrt{L_1 \cdot L_2} \cdot i_1}{1 + (j\omega L_2 + R_2) \cdot \left(\frac{1}{R_L} + j\omega C_{ges}\right)} \tag{3.18}$$

with $C_{ges} = C_1 + C_{tun}$ being the total capacity of the second side of the system including parasitic capacitance C_1 mainly caused by the inter-turn capacitance of the antenna. The maximum induced voltage is achieved when the system works at the system's resonance frequency.

Figure 3.2 shows a schematic of the coupled resonance circuit containing the primary side antenna L_1, the secondary side antenna L_2, the induced current i_2, the parasitic capacitance of the coil winding and the parasitic resistance of the antenna winding as well as of the connections toward the rest of the components. The other capacitance C_{tun} is the intentionally added tuning capacity to obtain a resonant circuit.

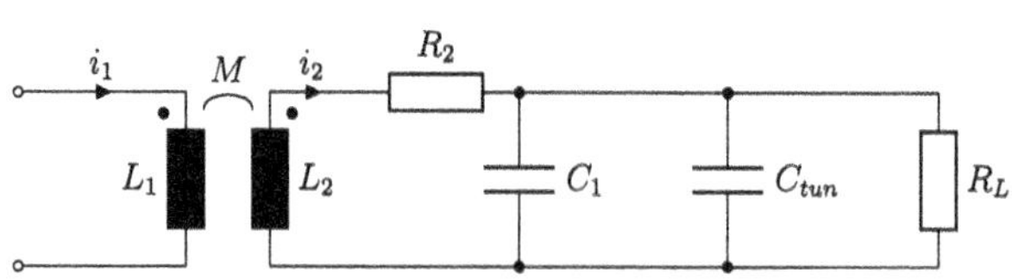

Fig. 3.2 Circuit diagram of the transponder IC with tag antenna (L_2) and primary side antenna (L_1)

The resistance R_L represents the load applied to it, which can be seen as the energy harvested to supply the sensor system. The more power is required by the system, the smaller the theoretically present R_L gets.

3.2.5 *Calculations for Planar Coil Geometries*

The inductance of a planar coil can be calculated using different approaches. For this work, the modified Wheeler's equations, provided in [28] are used for the layout. They are based on the original Wheeler's equations from 1928 [29]. In dependence on the geometry, the equations provided differ slightly which is expressed by the coefficients K_1 and K_2 in

$$L_{mw} = K_1 \cdot \mu_0 \frac{N^2 \cdot d_{avg}}{1 + K_2 \cdot \rho_f}, \tag{3.19}$$

valid for rectangular, octagonal, and hexagonal planar coil inductance with d_{avg} being the average diameter

$$d_{avg} = \frac{d_{out} + d_{in}}{2} \tag{3.20}$$

and ρ_f being the fill ratio

$$\rho_f = \frac{d_{out} - d_{in}}{d_{out} + d_{in}}. \tag{3.21}$$

The coefficient values can be found in Table 3.1.

For the circular design, the following equation can be used to calculate the inductance [30].

$$L_{Circ} = 31.33 \cdot \mu_0 \cdot N^2 \cdot \frac{(r_{in} + r_{out})^2}{16 \cdot (r_{in} + r_{out}) + 44 \cdot (r_{out} - r_{in})} \tag{3.22}$$

Table 3.1 Coefficients for modified Wheeler equations

Layout	K_1	K_2
Square	2.34	2.75
Hexagonal	2.33	3.82
Octagonal	2.25	3.55

3.2.6 Resonance Frequency and the Quality Factor

For a system having two separate sides that are required to work together, as in the case of resonant inductive coupling, the achievement of the targeted resonance frequency can be declared as one performance criterion for the antennas. Another important factor is the quality factor or Q-factor in this context. Resonance frequency as well as Q-factor can be determined from measurement data of the impedance of an LC-circuit. Therefore impedance becomes an additional parameter of interest for HF-RFID antennas.

Impedance (Z) is known as a measure of the resistance a circuit offers to an Alternating Current (AC) signal. It is a frequency-dependent parameter of a circuit and split into a real (resistance R) and an imaginary (reactance X) part, with the impedance being represented in different forms

$$Z = R + jX = |Z| \cdot e^{j\phi} = |Z|\underline{/\Theta} \tag{3.23}$$

and the absolute amount of impedance is calculated by

$$|Z| = \sqrt{R^2 + X^2}. \tag{3.24}$$

For theoretical elements, reactance is represented by capacitors and inductors and can be calculated using

$$X_C = \frac{1}{2\pi f C} \tag{3.25}$$

and

$$X_L = 2\pi f L, \tag{3.26}$$

respectively. In theory, for an ideal parallel LC-resonance circuit, the real part of the impedance is zero. The imaginary part, however, varies over frequency. From Eqs. 3.25 and 3.26, it can be seen that with increasing frequency, the reactance of the capacitor will decrease, while the reactance of the inductor will increase. Resonance occurs when the following condition applies:

$$X_L = X_C \tag{3.27}$$

This is the case at a frequency of

$$f_r = \frac{1}{2\pi\sqrt{LC}} \tag{3.28}$$

which is called resonance frequency. The impedance at this point is infinite for theoretical LC elements, while it is purely defined by the resistive share of the LC circuit and the used measurement setup for real circuits.

The Q-factor is closely connected to the resonance frequency and to the power transfer from one antenna to the other. It is a measure for the damping of an oscillator, meaning that mainly reactive component influences and less resistive ones cause a high Q-factor. For RFID tag resonance circuits, the Q-factor is defined as the ratio of resonance frequency to the −3 dB bandwidth [31]

$$Q = \frac{f_R}{f_{h,-3dB} - f_{l,-3dB}} \tag{3.29}$$

with $f_{h,-3dB}$ being the higher frequency where the signal shows −3 dB of its maximum amplitude and $f_{l,-3dB}$ being the lower frequency showing this amplitude. A higher Q-factor is determined by a very high increase of impedance around the resonance frequency, combined with reduced bandwidth and vice versa. That means that energy transmissions between two devices with a very high Q-factor are very efficient, but it also means that if one of the systems is slightly detuned, the efficiency gets very low very quickly. Therefore, the Q-factor in RFID systems is always a trade-off between good power transmission and thereby mostly longer distances and sensitivity against environmental influences occurring during operation. Especially for reader antennas, a lower Q-factor is usually targeted to allow for a larger bandwidth that is covered. For the tag antenna, the desired Q-factor depends strongly on the application. For applications using the ISO/IEC 14443 standard, a Q-factor between 7 and 10 is recommended, while for ISO/IEC 15693 applications an even higher Q-factor of up to 18 can be chosen to achieve higher read range [32].

3.2.7 High Permeability Materials

Electrically conductive materials around an antenna can lead to a massive performance decrease up to a complete failure of power transfer and communication. The reason for this can be found in eddy currents which are induced into the conductive material by the reader field following Faraday's law, mathematically described in one of Maxwell's Equations 3.1. These currents lead to a magnetic field on their own after the second Eq. 3.2, which is counteracting the change of flux of the primary field following Lenz's law. The amount of eddy currents induced is proportional to the conductivity of the material. Therefore, good conductors like copper or aluminum result in a higher influence when placed in the proximity of an RFID antenna than, e.g., unidirectional CFRP. Nevertheless, whenever a conductor is present in close proximity to an antenna, the field will be attenuated by some amount and thereby the coupling factor will be reduced.

At the same time, many applications demand for antennas working close to metal or other conductors, e.g., the RFID antenna in smartphones is usually directly placed on the back side, directly on the battery. For many applications, the most simple solution is to increase the distance between conductive surfaces and the antenna as far as possible. This approach shows results with only a few centimeters of space in between. Even if it is not possible to increase the distance in a way that eliminates the negative influence of the conductor, as soon as the influence is small enough, the remaining frequency shift can be retuned with the help of an additional tuning circuit.

For some applications, however, increasing the distance between conductive surfaces and the antenna is not an option. An alternative can be found in materials having a high relative permeability. The idea is to prevent the occurrence of eddy currents in the underlying layers to a major part by concentrating the magnetic flux inside the high permeability layer and preventing most of it from entering the conductive layer.

The permeability μ of a material is given by

$$\mu = \mu_r \cdot \mu_0, \tag{3.30}$$

where μ_r is the relative permeability of the material and μ_0 is the magnetic field constant, $1.257 \times 10^{-6}\,\mathrm{V\,s\,A^{-1}\,m^{-1}}$. For vacuum and air, $\mu_r = 1$, meaning they are magnetically neutral. For paramagnetic materials like aluminum or platinum, with $\mu > 1$, a slight amplification of the magnetic field is observable, that is due to their natural dipole moment that leads to an orientation in accordance to the applied magnetic field [18]. The thermal movement however leads to a quick return toward their initial random alignment.

In ferromagnetic materials there are also natural dipole moments present, but in addition to that larger groups of dipoles tend to be oriented in the same direction due to a coupling of spins of neighboring atoms. These are called magnetic domains and are bordered by so-called Bloch walls, where differently oriented dipoles are present directly next to each other. When a magnetic field is applied, depending on the field strength, the differently oriented domains change their orientation according to the magnetic field direction. At higher frequencies, however, metallic ferromagnetic materials show less desirable behavior due to increasing eddy currents. Therefore, ceramic ferrites which show a similar behavior can be used instead. The corresponding effect is called ferrimagnetism. In ferrimagnetic materials, atoms with opposing magnetic moments of unequal magnitude are present. Both ferromagnetic and ferrimagnetic materials are defined by the characteristic that $\mu_r >> 1$. For magnetically linear, isotropic materials, a scalar representation of its permeability is sufficient.

For ferrites, however, it is important to specify μ_r as a complex value, which even varies over frequency (see Fig. 3.3). Whereas μ_r' represents the real part of permeability or its ability to redirect the magnetic flux and μ_r'' describes the imaginary part which is responsible for heat loss during a change of flux direction. Thereby, depending on the application, a low μ_r'' at the targeted working frequency can be important. At the same time, ferrite sheets have a high electrical resistance, preventing the occurrence of eddy currents inside them.

As ferrite sheets show a very high permeability, at the ferrite-air interface there is an abrupt change of permeability present. Similar to light refraction at the interface of air and water, magnetic refraction can be observed for the interface of air and the ferrite sheet due to the change of permeability. Conditioned by the third Maxwell's Equation 3.3 that defines the flux density as source-free, thereby the normal compo-

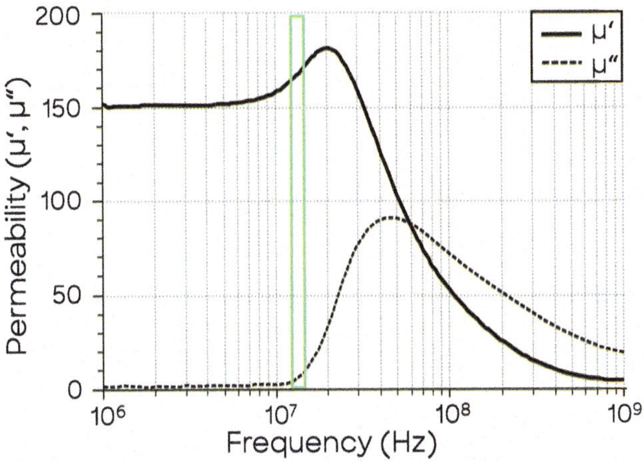

Fig. 3.3 Permeability of the used ferrite sheet EM15TF-014-1 from *3M* with marked working area (green), adapted from [33]

nent of the magnetic flux density has to be continuous. The normal component of the magnetic field however changes accordingly to the inverse ratio of permeability [18]

$$\frac{H_{n,1}}{H_{n,2}} = \frac{\mu_{r2}}{\mu_{r1}} \tag{3.31}$$

with $H_{n,1}$ and $H_{n,2}$ being the normal component of material 1 and material 2, respectively and μ_{r1} and μ_{r2} being the respective material's relative permeability. For the tangential components, the magnetic field shows a continuity (with an offset if an antenna coil is present nearby) while the flux density shows a direct dependence on the permeability ratio [18]:

$$\frac{B_{t,1}}{B_{t,2}} = \frac{\mu_{r1}}{\mu_{r2}}. \tag{3.32}$$

Combining these equations the magnetic refraction results to [18]

$$\frac{tan(\alpha_1)}{tan(\alpha_2)} = \frac{B_{t,1}}{B_{t,2}} = \frac{H_{n,2}}{H_{n,1}} = \frac{\mu_{r1}}{\mu_{r2}}. \tag{3.33}$$

For two materials with a large difference in permeability like air ($\mu_r = 1$) and a ferrite sheet ($\mu_r > 150$), this leads to $\mu_{r2} >> \mu_{r1} = 1$ which leads to $B_{t,2} = B_{t,1} \cdot \mu_{r2}$ and thereby to $B_{t,2} >> B_{t,1}$ this means a high concentration of magnetic flux inside the high permeability material occurs. This prevents a large part of the magnetic field, excited by the reader antenna, from entering the underlying conductive material and producing eddy currents. This prevents the negative effect of conductive layers underneath the ferrite sheet on the antenna performance, given either a sufficiently high permeability or thickness of the high-permeability layer.

For the design process, it has to be kept in mind that a ferrite sheet inserted directly under an RFID antenna increases the inductance of the coil. This happens because of the combination of Eq. 3.12 for the concatenated flux and Eq. 3.13 for inductance, which define the inductance as directly related to μ of the surrounding material.

3.2.8 Measurement Methods

For measuring impedance, there is more than one possible way to do this. From a measurement point of view, the auto-balancing bridge method is the best one suited for measurements at frequencies up to 120 MHz, for frequencies from 100 MHz to 3 GHz, the RF-IV method gives the best results, while for measurements at even higher frequencies the network analysis method is recommended [21]. As the target frequency of the tag antenna in this work is 13.56 MHz, the auto-balancing bridge method will be used to measure the achieved resonance frequency and quality factor with the help of an impedance measurement over a frequency sweep under different conditions.

Auto-Balancing Bridge Method

In simple theory, for the auto-balancing bridge method, a voltage is applied to the RFID tag antenna which is measured at the same time, while the current going through the device is also measured simultaneously. For the realization of this, no actual volt- or ampere-meters are used but the name giving bridge balancing. The measurement instrument balances a range resistor current with the Device Under Test (DUT) current with a voltage source at one side and a feedback path controlled second source at the other side, keeping the ground terminal in between at zero and monitoring this with a null detector. If the current through the range resistor is not the same as the DUT current, a balancing current will occur in the path to ground and lead the null detector to adjust the magnitude and phase angle of the feedback source, so that it becomes zero again. For frequencies above 100 kHz this is done by phase detectors in series with the null detector, separating the current into 0° and 90° components, integrating them and feeding the compounded currents back through the range resistor, thereby creating a control loop that is balancing the bridge. The vector voltages at two points across the bridge are then measured and used to calculate the complex Impedance of the DUT [21].

3.3 Design Process of FML Antennas

The antenna design is done in iterative steps. At first, simple geometries are chosen to validate the literature-provided theoretical antenna equations introduced in Sect. 3.2.5. As Mohan et al. mention in [28], square spirals are popular because of the ease of their design. Therefore, the first design is also square-shaped. To further investigate the influence of conductive materials beneath as well as the influence of ferrite, geometries are designed and measurements with antennas placed on surfaces with different conductivity and permeability are presented. The design is then adapted toward a structure working in direct proximity of conductive CFRP retuned to the correct working frequency for the final application. Since this depends strongly on the desired application of the structural part designed and can vary massively within one system e.g., over the parts of a complete airplane, attention is given to provide a solution that is sufficiently independent from the underlying material configuration.

All stainless-steel antenna structures designed and used within this work are manufactured by the Institute of Micro Technology in Braunschweig using femtosecond laser structuring from a thin metal sheet. This is done to exploit the possibilities of working cooperations within the research unit. Therefore, geometries are drawn using AutoCAD as a closed structure and handed over in a digital format for manufacturing.

3.3.1 Materials for Laminate Structured Antennas

From the theoretical point of view, any conductive material can be used to create an antenna structure for RFID applications. The material choice is often based on application dependent preferences. Low specific resistivity of the conductor material is the main concern to achieve a good quality factor when fabricating the antenna. Especially silver and copper are very well suited for this purpose (see Table 3.2). For very small antennas this can make a difference and also on the reader side of the system, copper antennas are found very often. Nevertheless, for some applications, small material price and weight are of higher importance, especially for the tag side of the system. Therefore, many tag antennas in mass production applications are

Table 3.2 Electrical conductivity for different antenna materials

Material	Conductivity
Silver	$62.1 \times 10^6\,\mathrm{S\,m^{-1}}$
Copper	$58.7 \times 10^6\,\mathrm{S\,m^{-1}}$
Gold	$44.2 \times 10^6\,\mathrm{S\,m^{-1}}$
Aluminum	$36.9 \times 10^6\,\mathrm{S\,m^{-1}}$
Stainless steel 1.4310	$1.37 \times 10^6\,\mathrm{S\,m^{-1}}$

made from aluminum, being comparably cheap, lightweight and still providing good electrical conductivity. As a commonly used material in aircraft construction by itself and as the metallic part of the standardized and most commonly known FML type called GLARE, it represents an alternative material for the planar RFID antennas investigated in this work. Therefore, contacting approaches are also investigated for aluminum. The aluminum sheets used for this are 0.4 mm thick, which is much thicker than those antennas usually found in RFID applications. Nevertheless, due to expected corrosion effects on the interface, aluminum is not combined with CFRP layers and therefore conductive substrates are not that much of a problem in this case. The results achieved with the material used in this work are expected to be transferable to aluminum designs directly, even offering a better performance due to increased conductivity.

The mainly investigated antenna material will be stainless steel. It is chosen as a good match to be combined with CFRP on one hand and being the much more challenging material for an antenna due to its low conductivity compared to other conductors on the other. In combination with carbon fibers, stainless steel is to chose over aluminum because the latter is known to suffer from galvanic corrosion which leads to early adhesive failure and material strength decrease [34]. CFRP is also the more challenging FML type due to the presence of conductive carbon fibers directly beneath the antenna coil. Therefore, antennas made from stainless steel on CFRP substrates will be used in the investigations presented here.

3.3.2 Laser Structuring Approach

All antennas are produced from 100 µm thick stainless-steel sheets, using the femtosecond laser ablation process at the Institute of Microtechnology (IMT) in Braunschweig as a Research Group interconnection. The designs are therefore sent over digitally. All antennas are produced on a 80 µm thin, heat resistant [35] handling foil to prevent the structure from bending out of its form due to material intrinsic stress. The design is set up using the AUTODESK® AutoCAD® software. Closed polygons are defined to enable the femtosecond laser software to define closed shapes to cut along. One example of a structuring specification can be seen in Fig. 3.4a. The extraction structures, marked in the figure, are intended to enable a careful removal of the cut-out gap material between the coil windings. The contours achieved with this structuring method are rough (see Fig. 3.4b), but for structures the size of the ones used in this work, and even smaller designs, this approach works very well.

3.3.3 Contacting Approaches

Different methods to achieve a reliable electrical contact between the antenna and sensor node can be considered. In this section, their compliance with electrical and

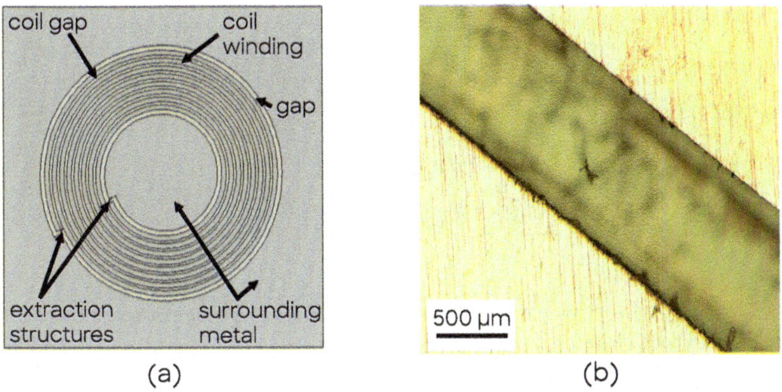

Fig. 3.4 Structuring of FML antennas. **a** CAD design for laser structuring. **b** Contour closeup for laser ablation structuring

mechanical requirements during measurements in a laboratory environment as well as the embedding process is investigated. Three possible approaches are investigated regarding their suitability for contacting the stainless steel used for the antennas within this work. To establish the connection for laboratory purposes, enameled copper wire with a copper diameter of 0.22 mm is mainly used for experimental setups. The same wire type with a copper diameter of 0.1 mm is used for some of the embedding experiments. Thinner wires turned out to be too thin and fragile for handling, especially during the embedding into larger plate geometries.

As one of the most commonly used methods for electrical contacting, soldering to both components of the system is investigated first of all. The first examined solder for this is *Stannol® Sn60Pb40*. On the sensor node part of the system, this approach clearly provides a very good electrical and satisfying mechanical contact. This can be expected when soldering to designated soldering pads on a PCB. When soldering to stainless steel using the *Sn60Pb40* solder, however, no proper substance-to-substance bond can be achieved. Even if a small amount of solder can be applied, the resulting bond is mechanically weak and can be scratched off easily with little force. Different soldering temperatures as well as heating of the whole work piece before soldering show no significant impact on the result. After this, another attempt of soldering directly to stainless steel is made using a specialized flux and solder combination for stainless steel with which it is in fact possible to establish a good soldering contact to stainless steel. The drawback of this method is, however, that the used flux is highly corrosive and has to be handled with care and completely removed carefully after soldering.

The second investigated contacting method, mainly known from the packaging of integrated circuits is wire bonding, where thin wires of gold, copper, or aluminum are pressed down and welded to the substrate using ultrasonic energy. The tested wedge bond (see Fig. 3.5a) can be safely achieved on stainless steel. The main drawback of this connection approach is the small diameter of the bond wire. With available

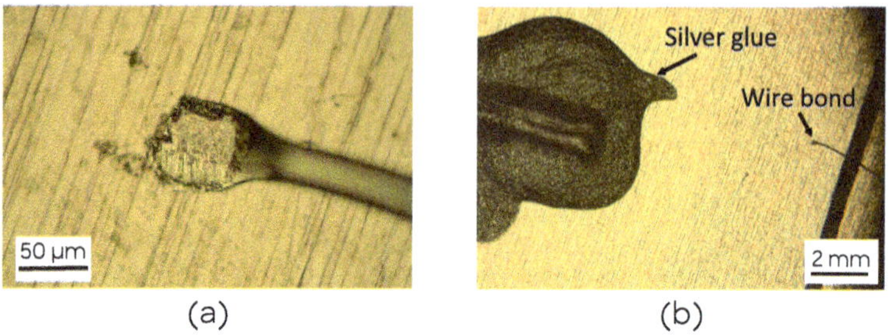

Fig. 3.5 Contacting methods for antennas. **a** Close-up image of a bond contact. **b** Comparison of silver glue and bond contact

machinery, only thin wires, normally used to connect a bare die to its surrounding package, could be tested. Despite providing excellent electrical contact, this results in a mechanically insufficient connection for the desired application. Nevertheless, there are possibilities to execute this exact process with significantly thicker bond wires up to 600 µm using a thick-wire bonder. Therefore, also not used within the frame of this work, this contacting method can be considered for similar future applications, given the availability of such a machine.

The third contacting method under investigation is the use of electrically conductive glue. *Panacol Elecolit® 414*, a glue containing silver particles, is applied to the contacting area and is hardened in the oven at 70 °C for 16 min. After additional 24 h, the glue is completely solid, and very good electrical contact is achieved. For antenna structures with dimensions comparable to the ones used also the comparably large contact area compared to, e.g., bonding (see Fig. 3.5b) is acceptable. The main drawback of this method, the high brittleness of the solid glue, can be dealt with by applying an additional layer of Ultra Violet (UV) curing glue type *DELO® KATIOBOND® GE680* after the conductive glue has fully solidified. Due to the straightforward application and reliable electrical connection achieved, this method is the chosen one that is used for contacting the manufactured antenna structures for all measurements done within this work.

3.3.4 Design Method Verification with Simple Designs

The recommended approach for RFID tag antenna designs is to use the equations given in Sect. 3.2 to find a geometry that fits the required inductance as a complementary part of the planned LC-resonance circuit and to manufacture three prototypes. One with exactly the theoretically required inductance and two others with 5 % more and less inductance [30]. Resonance frequency measurements are then performed to identify the best match among these three antennas.

Fig. 3.6 First antenna designs for verification of given equations. **a** Commercial tag with antenna for comparison [36]. **b** Rectangular antenna design. **c** Octagonal antenna design

The first antennas are of square and octagonal shape (see Fig. 3.6b and c, respectively), to have a fast and simple verification that the equations given in Sect. 3.2 are also valid for stainless steel antennas.

To keep the cut-out area small inside the FML components where the antenna is later cut into, the antennas in this project have a slightly different shape than usually seen RFID antennas (see Fig. 3.6a). The idea behind this is to increase the number of turns, thereby reducing the necessary outer diameter but at the same time increasing the turn width to reduce the overall coil resistance as far as possible. This leads to a significantly increased difference between outer and inner diameter, but an overall smaller antenna.

The design is done using Eq. 3.28, converted to

$$L = \frac{1}{4\pi^2 f_r^2 C_{tun}} \tag{3.34}$$

with the tuning capacitance inside the RFID Integrated Circuit (IC), which is measured to be 28.85 pF. With this, the targeted inductance value is 4.78 µH. As the design guide recommends, one version with a 5 % lower inductance of 4.54 µH and one with a 5 % higher inductance value 5.02 µH is manufactured to see which layout fits best to the desired application.

The measurements for the octagonal approach show that the calculation for the inductance value brings the antenna into the correct area, but the achieved resonance frequency for both 5 % versions is still below the targeted frequency of 13.56 MHz (see Fig. 3.7). Rey et al. report in [37], that the inter-winding parasitic capacity acts in parallel to the mutual inductance of the coil. Therefore, this value will be added to the parallel tuning capacitance, increasing the total value. It is assumed, that due to the different design approach compared to other RFID tag antennas by using more windings, there are higher parasitic capacities which add to the tuning capacitance and thereby increase the denominator of Eq. 3.28.

The phase progressions for both antennas are also plotted. Here it can be seen that as expected, the phase shifts at resonance frequency from a positive to a negative phase angle. This is the expected behavior for LC-resonance circuits and will be

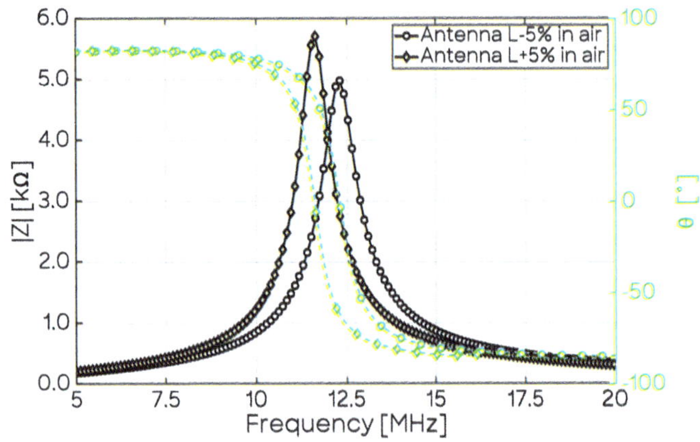

Fig. 3.7 Measurement of the octagonal antennas designed with +5% and −5% inductance. Dark lines: impedance. Green lines: phase

happening for all other measurements presented in the following. Therefore the phase progression will not be shown in addition to the impedance in the following plots for resonance frequency determination.

The results show that calculations are therefore in general still valid for rectangular and octagonal stainless steel antennas although they are made from a significantly worse conductor material. Nevertheless, the following designs are done in a way that there are fewer windings necessary, thereby also decreasing the coil resistance.

3.3.5 Geometrical Considerations

RFID antennas come in almost all imaginable simple geometries. Round ones are usually found in RFID key fobs, while for other applications, e.g., credit cards often rectangular shapes are considered. It can be seen from the design formulas in Sect. 3.2.5, that the geometric choice of the antenna has only a small effect on the final inductance. Two other requirements will influence the final antenna geometry decision by far more significantly. The antenna will be placed on a piece of conductive material as explained in Sect. 1.1. This substrate is not only conductive but the conductivity is anisotropic, i.e., it depends on the current path through the material. Although using an antenna with a high length-width aspect ratio may lead to advantages when used on unidirectional fiber configurations, it also limits its use in later applications. Due to a reduced influence on two sides, the layer configuration of the material varies from part to part, therefore making it necessary to choose the right antenna in the right orientation for each and every single part. To liberate the system of any alignment restrictions in the later application, a symmetrical design will be preferred over an asymmetric one.

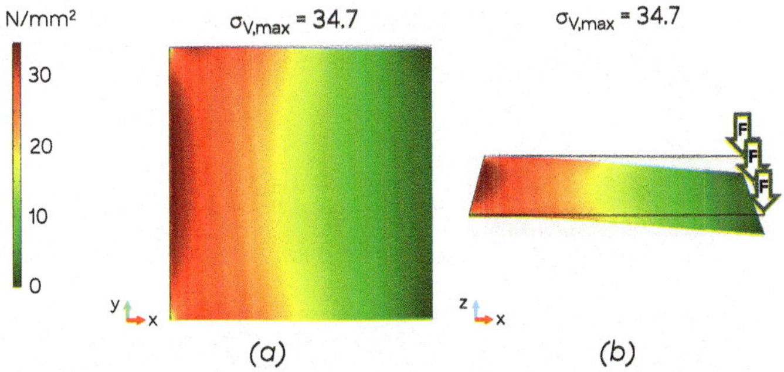

Fig. 3.8 Simulation of von Mises stress (N/mm^2) on a 120 µm thick stainless steel plate, fixated at the left side of the plate and a force of 67 mN applied to the right side's edge. **a** Top view of the plate. **b** Side view of the plate with the direction of force marked

The second requirement that significantly influences the geometry decision is the mechanical effect on the rest of the FML structure. The components manufactured with a lasered antenna as a defect in their top material layer will often be used for structural parts. This means they are experiencing massive loads in service. Therefore, the influence on the mechanical properties of the material has to be kept as small as possible. To identify the least weakening geometry to be used for the antenna a mechanical simulation is set up.

The simulated environment consists of a square plate with a width of 100 mm made from the top layer material, i.e., one layer of 120 µm thick stainless steel (see Fig. 3.8a). This is in compliance with the later on manufactured test specimens that are used to perform the impedance measurements. The plate is fixed at one edge and bent down using a specified amount of force on the opposite edge (Fig. 3.8b). The von Mises stress is chosen as the interesting parameter in this simulation because it offers a scalar value of stress that can then be used to estimate the maximum point up to which stress causes elastic deformations of the material. If the von Mises stress value exceeds the yield strength, provided in the material's datasheet, one can assume that a plastic deformation of the material will occur.

Adding a cutout into the plate, with a shape depending on the antenna geometry (see Fig. 3.9a–f) however, introduces some points to the structure which suffer from increased mechanical stress. Therefore, the force acting on the right side of the plate is increased iteratively up to 67 mN. This is the point where one of the simulated plates shows a maximum value slightly above the 0.2% proof strength provided for stainless-steel 1.4310 (X10CrNi18-8), namely 195 N/mm^2. For a layer containing no gaps, this leads to a maximum von Mises stress of just 34.7 N/mm^2. For the rectangular geometry, this force already causes a maximum von Mises stress of 200 N/mm^2 according to simulation results. It appears at one of the corners of the cutout (see Fig. 3.9a and d), making it the least suited choice for a mechanically compliant design of the antenna geometry. The simulation for an octagonal shape of the cutout (see Fig. 3.9b and e), results in a significantly lower maximum stress of

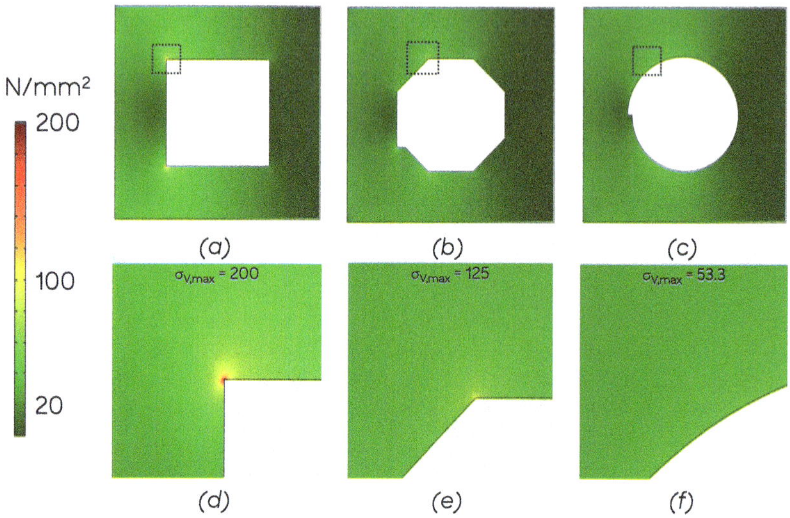

Fig. 3.9 Simulation of von Mises stress (N/mm^2) plates with antenna fabrication induced cutouts, fixated left side of plate, force of 67 mN applied to right side's edge. **a** Rectangular. **b** Octagonal. **c** Round. **d** Zoom in maximum stress area of rectangular geometry. **e** Zoom in maximum stress area of octagonal geometry. **f** Zoom in maximum stress area of round geometry

125 N/mm^2, while the best geometry according to the simulation results is the round approach (see Fig. 3.9c and f), with a maximum stress of 53.3 N/mm^2.

The round approach results in only 27% of the peak stress that would be caused if a rectangular antenna geometry would be chosen to be cut into the material. Given the results of this simulation, for the following investigations, a circular approach is preferred over other geometries.

3.4 Measurements

Within this section, the developed antenna structure will be validated by measurements. If not mentioned differently, the antennas are manufactured using square metal sheets of 4900 mm^2 with $N = 8$ turns, a conductor width of w=1 mm, an inner diameter of $d_{in} = 24.4$ mm, an outer diameter of $d_{out} = 51.6$ mm and thereby a theoretical inductance of $L_{theo} = 3.01\,\mu$H. The metal sheets are attached to a 80 µm thin handling tape before the antenna is cut to prevent deformation. As this goes for all experiments presented in this section, the term antenna will always apply to an antenna on its handling tape.

3.4.1 *Small Gaps and Surrounding Metal*

The antennas in the following are manufactured in a way that they keep the inner and outer metal areas present. This is due to the intended manufacturing of antenna coils directly into the material in the later application, where the surrounding material will remain as well. One requirement to the system design is the maximum interstice in the target material of 1 mm, which is supposed to be sufficiently small to not seriously degrade the metal layer properties. All antennas show a gap size of less than 1 mm over the whole design while keeping the distance to the outer and inner metal areas close to 1 mm to keep the influence on the antenna resonance frequency as small as possible. The influence of the surrounding metal areas on the resonance frequency and on the achieved quality factor is measured (see Fig. 3.10b).

With only the inner metal area added to the coil, the resonance frequency is shifted from 15.46 to 16.48 MHz, while the maximum impedance reduces and the bandwidth slightly increases. This means the quality factor which is determined using Eq. 3.29 decreases from $Q = 8.4$ for the antenna without metal surroundings to $Q = 6.81$ with inserted inner metal area. The influence is even more significant when the outer metal area is inserted in addition to the inner one. The resonance frequency is shifted to an even higher frequency of 17.83 MHz and the quality factor decreases to only $Q = 5.47$. It can be noticed that it is even more important to keep the coil resistance as low as possible to still be able to reach a sufficiently high quality factor and thereby have a system with a good power reception capability over a longer range.

3.4.2 *Measurement Setup: Clamping Fixture*

In the following experiments, the antennas are placed on layers of CFRP and ferrite sheets. To achieve a comparable material stack to an antenna structured after FML plate manufacturing, the layers have to be pressed onto each other and held

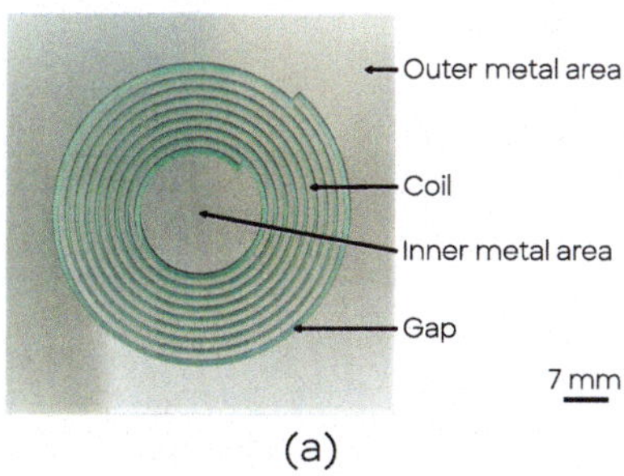

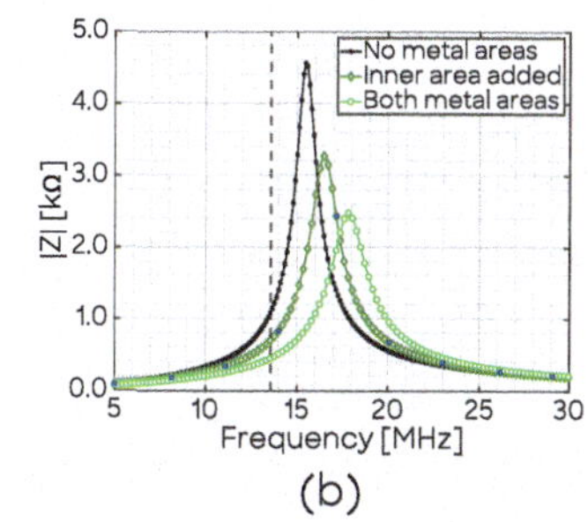

Fig. 3.10 Influence of metallic areas around the antenna coil on impedance over frequency. **a** Photograph of antenna with definition of surrounding metal areas. **b** Comparison of antenna impedance without any metallic surrounding to middle area inserted to complete surrounding areas present

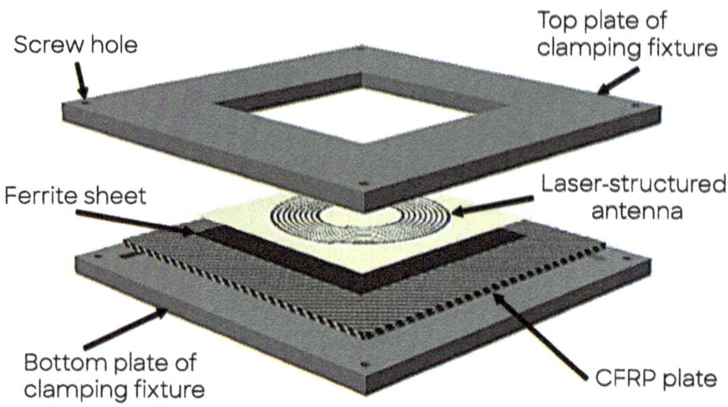

Fig. 3.11 Exploded-view drawing of the clamping fixture

in place during measurements. In addition to that, the Antenna structures on top of the handling foils are not completely flat so not all areas would have the same distance from the underlying material without tying them down. For this purpose, two 3D-printed clamping fixtures are designed, one for the smaller and one for the larger metal sheets. By using a bolt-nut combination in each corner, sufficient fixation can be achieved for the measurements. Figure 3.11 shows an exploded-view drawing of the clamping fixture used with all layers of the material stack inside.

3.4.3 Antennas on Different Ferrites

For this experiment, the medium antenna is placed on top of different types of ferrites which can be found in Table 3.3.

The results show that the combination of increased thickness and increased real part of relative permeability decreases the achieved resonance frequency (see Fig. 3.12). This is expected due to Eqs. 3.12 and 3.13 which relate the inductance of

Table 3.3 Ferrite sheets used for the experiment. μ_r' and μ_r'' given at 13.56 MHz, thickness given as thickness of ferrite without other layers and enumeration for Fig. 3.12.

No.	Type	Manuf.	Application	th [mm]	μ_r'	μ_r''
1	FFSX-0.3H-5060T	Kitagawa	HF-RFID	0.3	90	1
2	FFSW-0.1-5060T	Kitagawa	100 kHz	0.1	250	300
3	364 103	Würth	HF-RFID	0.24	150	3
4	364 101 4	Würth	HF-RFID	0.1	150	3
5	364 104	Würth	HF-RFID	0.34	150	3
6	EM15TF-014-1	3M	HF-RFID	0.12	168	5

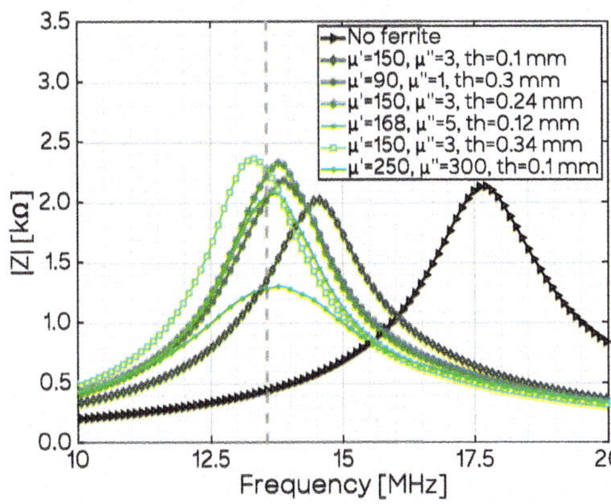

Fig. 3.12 Measurement of the medium antenna placed on different ferrite sheets, having different relative permeability values and thicknesses

a coil directly to its surrounding material. Thin ferrite sheets having a higher relative permeability value can result in the same resonance frequency as a thick ferrite sheet with less μ_r'.

The influence of a ferrite sheet designed for a different application, wireless charging at around 100 kHz, is also investigated. Besides reducing the resonance frequency, it reduces the achieved Q-factor in the measurement. This can be explained by an increased imaginary part of the relative permeability μ_r'' for this sheet at 13.56 MHz which causes increased heat loss due to a change of magnetic flux direction at this frequency.

Combined with CFRP, the total influence on the achieved resonance frequency is expected to vary with the type of CFRP used beneath, or in other words with the conductivity of the layer placed beneath. This will be investigated in the next step.

3.4.4 Antennas on Different CFRPs

As introduced in Chap. 2, there are three different layer stack configurations available for measurements in this project. The unidirectional stack-up having a comparably low electrical conductivity, than the perpendicular one with alternating layer orientations and the multidirectional one, having even more variety in its layer orientations. All three plates are expected to have a different influence on the achieved antenna resonance frequency if the antenna is placed in very close proximity to them. As a reference, the antenna measured inside the 3D printed clamping fixture made from PolyLactic Acid (PLA) (therefore named "PLA" in the legends) is depicted beside the measurements for the same antenna placed on different CFRP plates (see Fig. 3.13). It shows a resonance frequency of 18.28 MHz and a Q-factor of Q = 5.6.

It can be seen that for both perpendicular and multidirectional CFRP the influence is quite similar. Both are shifting the resonance frequency to a higher value and lower the Q-factor significantly. For the perpendicular stack up the resonance frequency

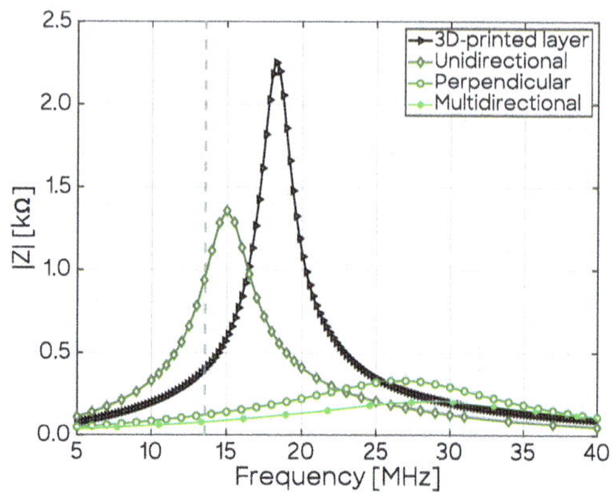

Fig. 3.13 Measurement of the medium sized antenna placed on different types of CFRP stack-up configurations and placed on 3D-printed clamping fixture for reference

shifts to 26.66 MHz with a Q-factor of Q = 1.42, for the multidirectional stack up to 28.4 MHz with an even more reduced Q-factor of only Q = 1.09. This is probably caused by the increased conductivity in three, instead of only two directions of the plate compared to the perpendicular stack-up.

The unidirectional stack-up on the other hand shows a reduced Q-factor of Q = 2.86 due to the conductive fibers inside as well, but also shows a slight decrease of the resonance frequency in contrast to the other two plates. It can be assumed that this may be caused by a higher permittivity of CFRP plates [38], compared to the measurement on 3D printed PLA. This results in a higher parasitic capacitance value, increasing the already present tuning capacitance by some pF and thereby shifting the resonance frequency to a lower value. With this, it would also be possible to simply adjust the design for the direct use on unidirectional CFRP of this fiber type and this fiber volume content, if the power transmission would be still sufficient despite a reduced Q-factor. Hence the aim is to identify a solution that is suitable to work for antennas placed on CFRP regardless of its fiber orientation to offer maximum placement and orientation flexibility, this option is not suited to be used within this project, though it might be a solution for special cases.

With this, the question arises if it is possible to tune the antennas back to their target frequency without still being dependent on their substrate. This will be investigated next.

3.4.5 *Antennas on Ferrite and Different CFRPs*

For the same plates used in the previous section, one of the ferrite sheets examined in Sect. 3.4.3 is chosen to be used. The choice goes for the *EM15TF-014-1* sintered ferrite sheet from manufacturer *3M*. It delivers the best combination of overall thickness and ferrite layer thickness combined with a high μ_r' value at 13.56 MHz. The results (see Fig. 3.14) show that the resulting resonance frequency, as well as the

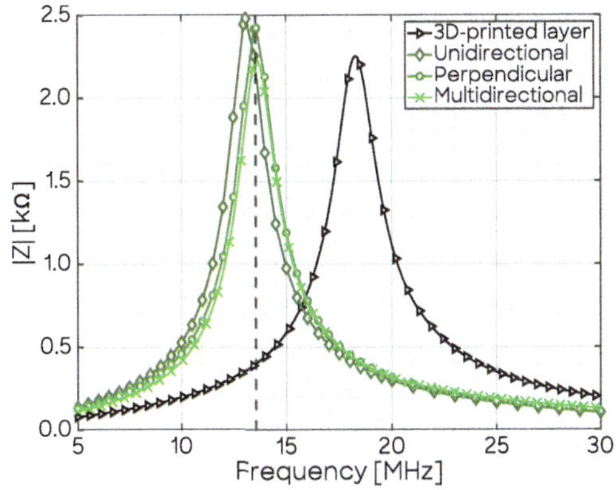

Fig. 3.14 Measurement of the medium sized antenna placed on different types of CFRP stack-up configurations with EM15TF-014-1 sintered ferrite sheet in between. Antenna on 3D-printed clamping fixture as reference

Q-factors, in fact, show an influence of the type of CFRP stack-up beneath them, but the influence is small and can even be neglected for a typical HF-RFID system using an off-the-shelf reader. These devices usually offer a lower Q-factor compared to the tags to ensure a resonance frequency match despite small environmental influences, therefore usually a wider range around 13.56 MHz. This way, tag resonance frequencies can vary to some amount without significant disadvantages.

The only thing left to examine is the influence of the embedding process on the ferrite sheets inside the material stack. If the behavior differs after they are exposed to 180 °C for several hours, this approach might not be suited after all. Fortunately, after putting the ferrite sheet into an oven for four hours at 185 °C, after a complete cooldown, no change in measurement results can be determined as compared to the ones done before the temperature was applied. Even the adhesive layer shows no noticeable degradation after this time. Therefore, the optimal setup to get planar spiral coil antennas working on conductive CFRP, when structured into the outer metal layer of FML, is found in a thin, high permeability layer of ferrite in between.

3.4.6 Designs for GFRP and CFRP Application

After the design for antennas working on conductive carbon fiber reinforced CFRP substrate, another design for antennas working on non-conductive substrates, e.g., on glass fiber reinforced GFRP is done. The resulting parameters for this version in comparison to the adjusted antenna for CFRP substrates are listed in Table 3.4.

The resulting resonance frequency of this system in air is measured to be at 13.66 MHz and a Q-factor of $Q = 4.35$ is achieved (see Fig. 3.15).

The decrease of the Q-factor compared to the other design can be traced back to a similar conductor width with at the same time increased coil length due to more windings to achieve the higher inductance value. Furthermore, it can be seen in Fig. 3.15 that in contrast to the non-conductive substrate approach, the design for

Table 3.4 Design parameters for the presented antennas

Parameter	Medium	GFRP
Number of turns (N)	8	9
Conductor width (w) [mm]	1	0.4
Inner diameter (d_{in}) [mm]	24.4	30.5
Outer diameter (d_{out}) [mm]	51.6	70.0
Theoretical inductance (L_{theo}) [μH]	3.01	4.81

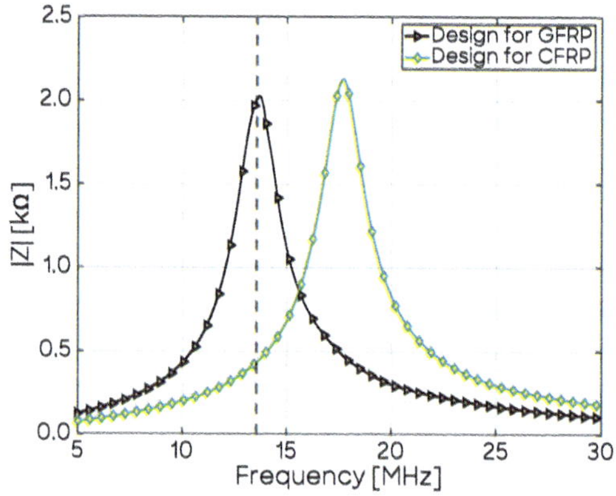

Fig. 3.15 Measurement impedance over frequency to determine the resonance frequency for antennas designed for GFRP substrate vs. those for CFRP substrate measured in air

conductive substrates is initially tuned 4 MHz higher because of the frequency shifts due to the combination of CFRP and ferrite beneath it, presented in Sect. 3.4.5.

3.5 Environmental Considerations

Airplanes, as one of the main end applications for the developed system, are exposed to comparably harsh environments. Given operation all over the globe, their exterior faces local temperatures and humidity levels. By cutting the antenna structure into the outer metallic layer of the FML, an unprotected zone is created where water can be absorbed by the underlying composite layers. Usually, the complete skin of the airplane is covered with a waterproof varnish. But nevertheless, the additional water absorption caused by unprotected antenna cuts will be evaluated in a laboratory setup here to see if different antenna cut configurations or the insertion of a ferrite layer cause differences in water absorption behavior.

Water Absorption due to Cutting Antenna Structures into FML

For the experimental setup, the idea is to take the unoccupied area from the large antenna structure and distribute this area over different cutout numbers and widths. The intention here is to simulate different antenna design approaches—increased diameter with reduced windings leading to fewer cuts, smaller diameter with more windings leading to more cuts following Eq. 3.22—on a comparable basis.

For this, thin aluminum metal sheets 45 mm by 90 mm are prepared with cutouts (see Fig. 3.16), an aluminum backplane without any cutouts and 8 layers of *UNIPREG* CFRP prepreg in between. The specimens are numbered to be able to assign the measurements done later on following the scheme x1 with one cutout, x2 with two, x3 with four, x4 with 8, x5 with 16, and x6 with 32 cuts, the "x" marking three specimens per type from 0 to 2.

The specimens x3 with four cuts are used to compare different deviating conditions: Specimen no. 03 has an additional layer of sintered ferrite inserted over its complete area, beneath the top metal layer with cutouts, to simulate an actual antenna stack for the use in combination with CFRP. Specimen no. 13 is sealed with spray-varnish of type *Kontakt Chemie—PLASTIK 70*, while specimen no. 23 is sealed with another type of spray-varnish *Kontakt Chemie—URETHAN 71*.

The edges are sealed with polyimide tape. As this is not a complete seal against humidity uptake, especially not over a longer period, an additional reference specimen is prepared consisting of two metal plates without cutouts and 8 layers of CFRP in between. With this, it is possible to capture the water uptake over the plate's edges simultaneously. Another test specimen consisting of 8 layers of pure CFRP is inserted as a reference as well to capture the humidity absorption behavior of the used type of CFRP.

The specimens are manufactured two days before the start of the climate chamber measurement, therefore the drying process before inserting them into the climate chamber can be decreased compared to similar climate experiments. The specimens are dried in an oven at 50 °C for 12 hours. After this time, the sealing spray coating is applied to specimens "13" and "23" and all specimens are weighed to record their initial weight. They are then inserted in a climate chamber configured to 50.5 °C at >80 % relative humidity. Temperature and humidity levels are monitored using a

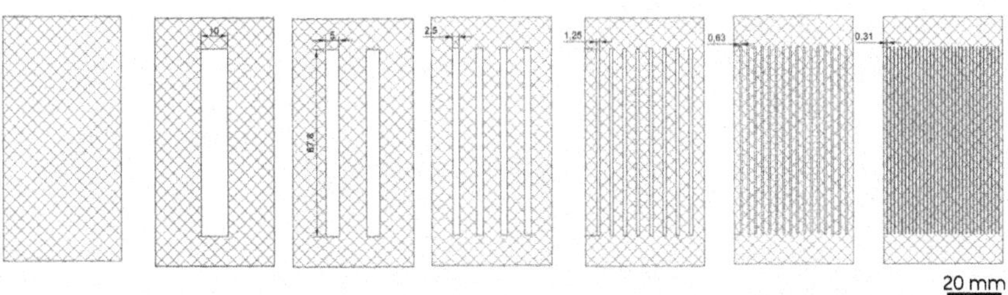

Fig. 3.16 Specimens for climate chamber experiment

combined temperature and humidity sensor. For measurements, the specimens are taken out of the climate chamber and set aside for about 10 min to cool down to room temperature. After that, they are weighed one by one using a scale with a resolution of 0.1mg.

It is expected that the x6 specimens with the highest amount of gaps will absorb more water in less time than the ones with fewer cuts and reach a saturated state first. At the same time, it is expected that the varnish coatings will prevent water absorption at least partly, while the specimen with an inserted ferrite layer is expected to either be protected from water absorption to a good amount by the additional polymer and adhesive layers or will absorb water even faster due to the porous volume of the ferrite layer itself.

The relative weight gain in percent is calculated according to

$$\Delta m_t(d) = \frac{m_t(d) - m_0}{m_0} \cdot 100\% \tag{3.35}$$

with $m_t(d)$ being the weight of the specimen at day d and m_0 being the weight of the dry specimen at the start of the experiment [39].

The results show that the expectations are only partly correct (see Fig. 3.17). The pure CFRP specimen's relative weight, represented by the blue triangles, increased

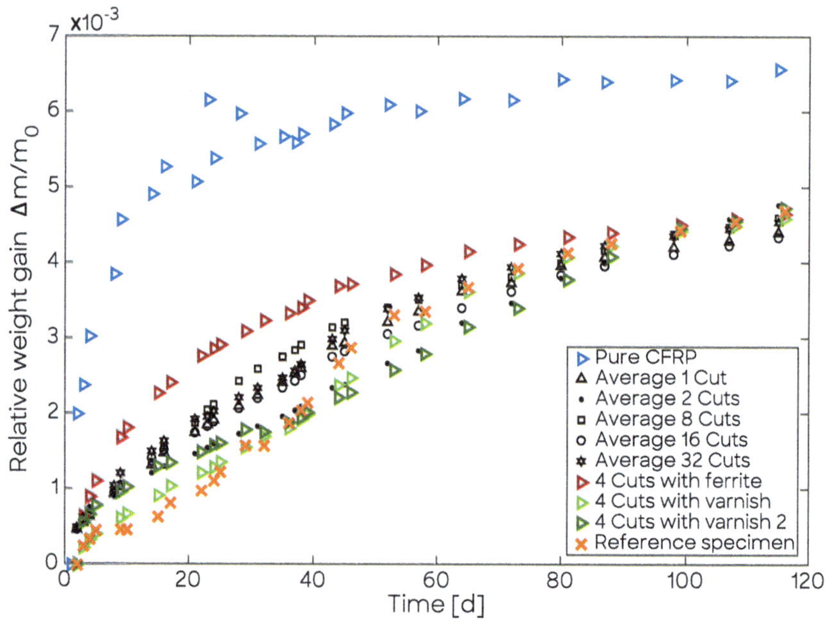

Fig. 3.17 Measurement results of climate chamber experiment: relative increase of specimen weight over days inside the climate chamber. Average values of three for specimens x1, x2, x4, x5 and x6

much faster than that of the metal-covered ones. This is exactly in accordance to the expected result: water is absorbed over much more surface area compared to the metal-covered specimens therefore it increases weight more rapidly and also approaches a state of saturation after about 80 days.

One reference specimen without any cutouts, represented with orange cross markers is inserted along with the other specimens. Comparably to the other metal specimens, it has the open edge area covered with one layer of adhesive tape. It represents the water uptake ratio over the edges for all specimens. It can be seen that the relative weight gain of this specimen is low compared to that of the other specimens for about one third of the total duration. After that, it rises to similar values as the unprotected specimens.

Both samples covered with spray varnish, specimen 13 with *PLASTIK 70*, depicted with light green triangles and specimen 23, covered with *URETHAN 71*, marked with dark green triangles, stay very close to the reference sample line. This means the varnish fulfilled its purpose in keeping the water from being absorbed. The absorption mainly happened via the open edges. They even show a lower absorption than the reference specimen after about one third of the total experiment duration. It is assumed that this is caused by the additional layer of varnish, better sealing the interface between metal and the edge covering adhesive tape compared to the other specimens.

Against expectations, the specimens with different amounts of cuts in their top layer show an almost linear increase of weight gain over the whole time, which seems to be overall independent of the number of cuts. Just the averaged x2 specimen line for samples with two cutouts, represented by black stars, shows a somehow reduced relative weight gain between days 14 and 72 while showing a comparable gain to the other samples at the beginning as well as at the end of the experiment. However, it is assumed that this is a statistical deviation and does not show an improved water uptake behavior for specimens with exactly two cuts in general.

For the specimen with a layer of ferrite inside, represented by the red triangles, a significantly higher relative weight gain can be observed. At the beginning of the experiment, the weight increase is much higher than for the other specimens. Toward the end of the experiment, however, this effect decreases again and the curve also adjusts to the other samples containing no ferrite layer inside. The assumption is that the sharper increase at the beginning may be caused by the high porosity of the ferrite layer itself, possibly influencing the water uptake from the side of the specimen.

The results of this experiment show that to maintain structural conformity of the airplane components into which antenna structures are cut, it is imperative to apply a covering layer of protective paint preventing water to get into the laminate. The need for this does not necessarily increase with the amounts of cutouts in the laminate but it certainly does increase if a layer of ferrite needed for the correct function of the antenna is inserted. As varnish is applied for pure CFRP and also for aluminum parts of an airplane anyways, it does not increase the effort in manufacturing when using FML with material-integrated antennas in comparison to other parts.

Regarding the low amount of identical specimens, especially for the varnished, ferrite, and reference specimens, as well as the improvable sealing of the edges, this experiment is only a preliminary test, which allows a first estimation of the water absorption behavior via the antenna structures. For a statistically more valuable statement, a repetition focusing on the most significant specimens of this test is highly recommended.

3.6 Antenna Diameter Influence

The antenna diameter has an influence on the inductance of the antenna as shown by Eq. 3.22. This can be taken care of by changing other parameters like the number of turns to get the same inductance value for two different designs. Nevertheless, changed conditions as for example the conductor width and overall length, will have an influence on the quality factor achieved.

Three different antenna sizes will be investigated here: a small antenna for better application compatibility, the medium-sized antenna used in the experiments presented before as a reference, and a large antenna structure to investigate possible advantages of this. The small and medium antenna structures are manufactured out of 4900 mm^2 square-shaped metal sheets while the large antennas are manufactured from 8100 mm^2 sheets due to their higher outer diameter. The geometrical parameters are listed in Table 3.5.

The main parameter that is expected to change using a larger antenna design is the maximum achievable harvested power by the antenna pair in use. The *ST25R3916-DISCO* RFID reader development board will be used for later sensor data readout. Therefore, the reader antenna diameter is fixed to 60 mm by 60 mm. Equation 3.18 shows that mainly coupling is responsible for a higher induced voltage on the secondary side. By increasing the outer diameter to 70 mm, the large tag antenna is much more in the range of the reader antenna, providing better coupling. In addition to good coupling, a reduced coil resistance increases the induced voltage. This can be realized when using an increased conductor width. Therefore, with higher power demands for the electronic sensor nodes, antennas large enough to provide a sufficient amount of coupling, small resistance and thereby increased power to the system will be required. Whereas for a system requiring less power to work sufficiently, also

Table 3.5 Design parameters for the differently sized antennas

Parameter	Small	Medium	Large
Number of turns (N)	11	8	7
Conductor width (w) [mm]	0.25	1	2
Gap to surrounding metal [mm]	1	1	1
Inner diameter (d_{in}) [mm]	13.0	24.4	31.7
Outer diameter (d_{out}) [mm]	25.0	51.6	70.0
Theoretical inductance (L_{theo}) [μH]	3.03	3.01	3.01

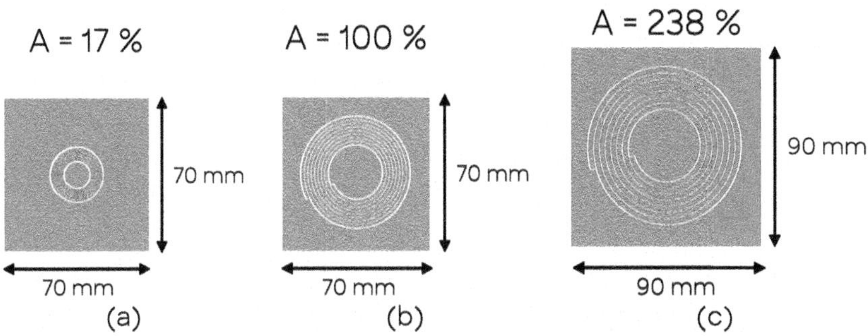

Fig. 3.18 Antenna geometries. **a** Small antenna. **b** Medium antenna. **c** Large antenna

a smaller antenna approach may be suitable to supply the system while keeping the influence on the FML smaller. The influence of different antenna diameters on the power supply capability of the wireless link is demonstrated in the energy harvesting measurements presented in the first part of this section.

One other correlation worth mentioning in the use case of this antenna is an increased influence of conductive and highly permeable materials close to the antenna as the antenna size increases. This will be discussed in the second part of this section.

The three different antennas manufactured for this are depicted in Fig. 3.18, the associated geometry details can be found in Table 3.5. The resulting coil area of the antennas is displayed in relation to the medium-sized antenna.

Energy Harvesting Capabilities of Different Antenna Sizes

The energy harvesting capability of differently sized antennas is expected to be different as well. If the reader antenna geometry remains the same, it is expected that the closer the tag antenna dimensions are to the reader antenna dimensions, the more effective the energy transfer will be. Using Eq. 3.9, the theoretically available magnetic field strength depending on the current through the reader antenna can be calculated at 10 mm distance to a square-shaped antenna coil with 60 mm length and two windings:

$$H(I) = 0.968\,\mathrm{A\,m^{-1}} \cdot I \tag{3.36}$$

A schematic measurement setup can be seen in Fig. 3.19. The energy harvesting output of the RFID IC is directly connected to a load resistance in parallel, over which the voltage drop is measured using a voltmeter. By this, it is possible to determine at which load and at which distance between the antennas the energy harvesting feature of the IC still delivers a sufficient supply voltage $>1.8\,\mathrm{V}$ for the sensor node to work.

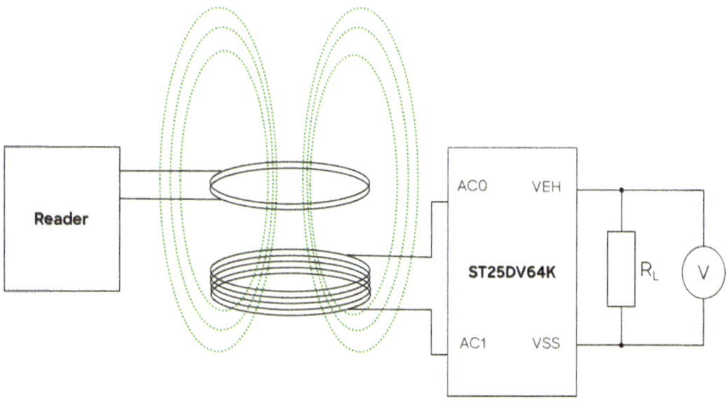

Fig. 3.19 Experimental setup for energy harvesting measurements

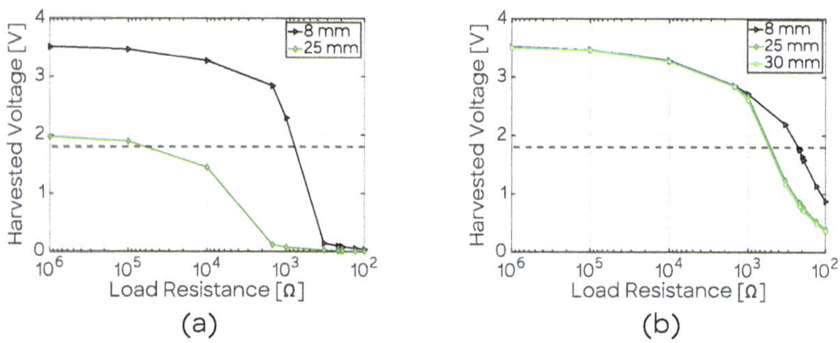

Fig. 3.20 Measurement of energy harvested as voltage over load resistance. **a** Small antenna. **b** Medium-sized antenna

As displayed in Fig. 3.20a, the voltage harvested with the small antenna stays above the threshold voltage for resistance values of 1 kΩ for an antenna distance of 8 mm. The voltage drop measured of 2.29 V equals 5.2 mW using Eq. 3.37.

$$P = \frac{U^2}{R} \tag{3.37}$$

For increased antenna distance, the voltage already drops below the threshold for values smaller than 100 kΩ. Therefore, in combination with the chosen reader antenna, the small antenna can only be used for sensor nodes which require very low amounts of energy to work.

The medium antenna drops below the voltage threshold for a load resistances of less than 329 Ω, equal to 14.5 mW at close distance. For a distance of 30 mm, this already happens for a load resistance below 1 kΩ, which is equal to 6.8 mW.

The results obtained for the large antenna show that it performs best in combination with the chosen reader antenna even over increased distances (see Fig. 3.21a). The second version with comparable outer diameter but reduced interstices and thereby

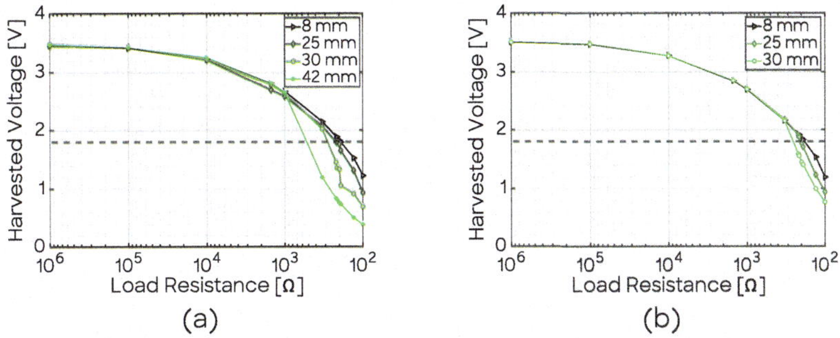

Fig. 3.21 Measurement of energy harvested as voltage over load resistance. **a** Large antenna. **b** Large antenna with higher fill factor

increased conductor width even outperforms the one with less conductor width at higher distances (see Fig. 3.21b). It is able to deliver a supply voltage of more than 1.8 V at 8 mm distance for a resistance of 191 Ω which means it is able to deliver up to 17.5 mW to a sensor node, given good antenna alignment. Nevertheless, to ensure a sufficient sensor node function even if the antennas are maybe slightly misaligned, or if the antennas are more than 8 mm apart, the sensor node power consumption should be targeted to stay below 15 mW.

Antenna Diameters and the Influence of CFRP and Ferrite

Another result obtained while measuring the characteristics of the antenna geometries with different sizes is a differently strong influence of CFRP on their performance (Fig. 3.22). It can be seen that the achieved resonance frequency for the three different geometries is not exactly the same even though they are all designed for a theoretical inductance of 3.0 μH. This can be reasoned with an increase of parallel capacitance due to increased coupling capacitance between the coils due to a higher number of turns. An additional characteristic is the reduced Q-factor of a smaller antenna compared to a bigger one. Two reasons for this can be determined: the increased number of turns, by which the overall conductor length increases, and by a reduced conductor width for the smaller antenna to decrease the outer diameter even more. Both changes lead to higher coil resistance.

Another thing that can be noticed is that the resonance frequency shift as well as the obtained attenuation of the antenna is significantly smaller for the small antenna than for the larger one. This is assumed to be justified by the measurement setup chosen to measure the input impedance of the antenna. For a higher antenna resistance, applying a constant voltage to it results in a smaller current, which, given Maxwell's second equation, results in less magnetic field strength, thereby causing a weaker opposing field caused by eddy currents in the CFRP. The influence of ferrite

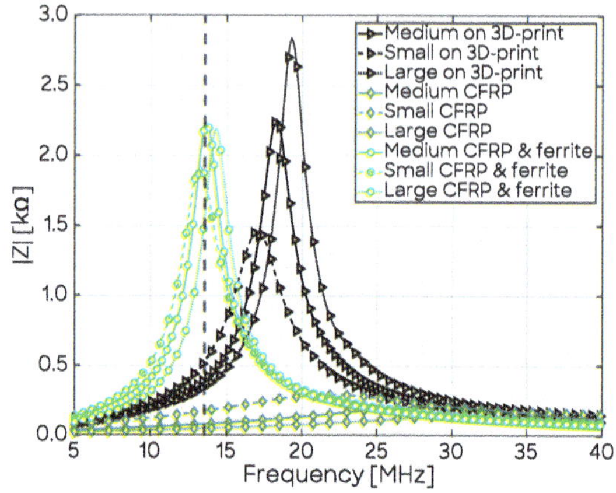

Fig. 3.22 Measurement of impedance over frequency for antennas of different sizes on 3D-printed PLA, CFRP, and with ferrite

regarding resonance frequency is not different between the antennas. As it increases the inductance of each antenna coil, the influence of parasitic capacitance between the windings on the resonance frequency is lowered by this, so they are moved closer together by this. For all antennas, the resonance frequencies are in good proximity to the target frequency of 13.56 MHz when a ferrite sheet is used.

What remains to be noticed, is the change of quality factor by the use of a ferrite sheet. For the small antenna, it is increased significantly, for the medium-sized antenna it stays in the same range and for the large antenna it is significantly reduced. It has to be kept in mind, whenever the antenna size is increased to yield a better coupling factor, that the quality factor in combination with ferrite is decreased by a noticeable amount.

3.7 Simulation Model for HF-RFID Antennas

The equations in Sect. 3.2 already depict quite vividly, which influence basically any change of and around the antenna can have on the inductance, the quality factor, and the resonance frequency achieved. Also, measurements on manufactured prototypes show all the effects. Although an estimation of the influence followed by an iterative design adaptation as demonstrated in the previous section leads to correctly tuned antennas for the application, this approach is time and material-consuming for possible changing boundary conditions in the future. Facing the variety of influences it is almost impossible to experimentally investigate them all. Thinking only of the importance of permeability and thickness combination in the used ferrite layers, multiple combinations are theoretically possible but some are simply not available for purchase at the moment of interest.

Therefore, a simplified simulation model is set up, using literature parameters for materials and verifying it with the results of experimental measurements with the aim to be able to estimate the influence of design decisions on the practical outcome before antenna manufacturing. Furthermore, with this also the influence of, e.g., different conductor materials and thicknesses can be investigated easily. The model setup is presented along with the simulation results for adjusted geometries and different materials around the antenna.

3.7.1 Simulation Setup

The simulation is done using the Finite Element Method (FEM) simulation software *COMSOL Multiphysics®*. With this, it is possible to combine different physical simulations which influence each other via defined interfaces. For the simulation of the complete material stack, an axis-symmetric, two-dimensional approach is chosen to be able to model all material layers in geometrical detail, without increasing the simulation model complexity too much. The two-dimensional axis-symmetric one offers the clear advantage of massively reduced simulation duration compared to a three-dimensional model.

Layer and Geometry Definition

As a first step, a symmetry axis through the model has to be chosen. The most accurate symmetry axis is assumed to go right through the middle of the design, leading toward the opposite side of the coil's start and end point (see Fig. 3.23a, green line). The thereby generated model is formed as multiple closed rings (see Fig. 3.23b). By choosing the RLC Coil Group option in the Magnetic and Electric Fields physics tool, the separate rings can be defined to form a coherent planar coil between the inside and the outside. The setup is placed inside an air-filled sphere. This sphere has an outer diameter that is 80 mm larger than the outer diameter of the simulated coil, adjusting to different coil diameters, to avoid any influences that come from the finite simulation space around the antenna. In addition to this, the outer 5 mm are defined as an infinite element domain, which means this region is simulated approximating an infinitely large domain.

Beneath the antenna coil itself, there are seven more thin layers defined, according to the experimental setup. The layer definitions like thickness, if not varied during simulation, and material parameters used can be found in Table 3.6.

Each layer can be activated and deactivated freely between simulations to get the simulation results for the later on performed experiments. The resonant circuit's input impedance is measured in dependence of its frequency on different materials like CFRP and ferrite sheets as already shown in the measurements in Sect. 3.4.

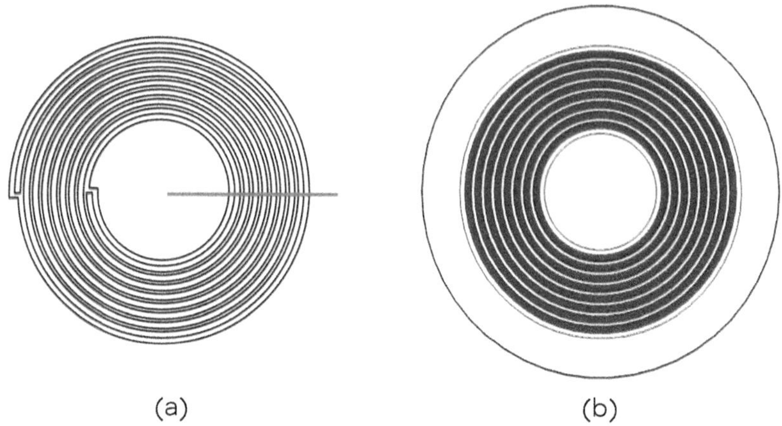

Fig. 3.23 Simulation model. **a** Cut-line through actual antenna geometry. **b** 3D view of the 2D-axis symmetric model in simulation

Table 3.6 Layers within the axis-symmetric simulation model. * complex permeability value at the operating frequency 13.56 MHz. ** Anisotropic value definition, see below

Layer name	σ [S/m]	μ_r	ϵ_r	Typ. thickness
Air	10e-10	1	1	r_{out}+80 mm
Stainless steel [40]	1.37e6	1.003	1	0.10 mm
Handling foil [35]	5e-14	2.1	2.65	0.08 mm
Cover Film (PET) [33]	1e-14	2.1	3	0.01 mm
Sintered Ferrite [33, 41]	1e-4	168-j5*	10	0.10 mm
Adhesive & release [33]	5e-14	2.1	2.65	0.09 mm
CFRP plate [38]	**	1	3.7	2.00 mm
3D-printed PLA [42]	5e-14	2.1	2.65	3.50 mm

The permeability of sintered ferrite is not constant over frequency as already shown in Fig. 3.3. Nevertheless, to keep the simulation model as simple as possible, it will be taken as a constant value extracted from the datasheet at the working frequency of 13.56 MHz.

For the definition of conductivity for the used CFRP, only one layer configuration is investigated in the simulation. The multidirectional laminate stacking sequence (see Table 2.1) is chosen as the CFRP layer in the simulation. It is one of the three small plates that is used for experimental validation in this work. To avoid tremendously increasing calculation expense by simulating each layer separately, not to mention the fibers inside epoxy resin separately, an estimated but still anisotropic conductivity is calculated for the three main material directions as follows:

One carbon fiber of type AS4 within each CFRP prepreg ply has a specified electrical resistance of $1.7 \times 10^{-3}\,\Omega$cm [43] and thereby a conductivity of $\sigma_{fiber} =$

$58.8 \times 10^3\,\mathrm{S\,m^{-1}}$. Given a fiber volume content in the cured material of 57.42% [44] and a conductivity of the surrounding epoxy matrix of $\sigma_{epoxy} = 10^{-9}\,\mathrm{S\,m^{-1}}$, this makes a conductivity of

$$\sigma_x = 0.5742 \cdot \sigma_{fiber} + (1 - 0.5742) \cdot \sigma_{epoxy} = 33{,}763\,\mathrm{S\,m^{-1}} \tag{3.38}$$

for each layer in fiber direction. According to the stacking sequence, all layers facing in x-direction (freely defined at this point) are containing fibers facing in this direction. Therefore σ_x is filled into the x-component of the conductivity matrix.

Assuming there is only epoxy resin conductivity perpendicular to fiber direction and taking the layer stacking setup into account once again, the conductivity in x-direction can be calculated. There are eight layers (all +/-45°) of a total of 13 layers containing conducting components in that direction. This makes up 62% of that direction's conductive layers. This leads to a y-component of the conductivity matrix of

$$\sigma_y = 0.62 \cdot \sigma_x + (1 - 0.62) \cdot \sigma_{epoxy} = 20{,}933\,\mathrm{S\,m^{-1}}. \tag{3.39}$$

In z-direction, assuming an ideal prepreg setup, $\sigma_x = \sigma_{epoxy}$ is taken into account.

3.7.2 Simulation Results

At first, the achieved simulation results based on the medium-sized antenna are compared with experimental measurements for validation of the developed simulation model (see Fig. 3.24). It can be seen that although the simulation result for the antenna on PLA shows a slightly higher input impedance value and thereby a slightly higher quality factor, the achieved resonance frequency is the same. Also, the shift in resonance frequency due to underlying CFRP matches with the measurement result quite well. In fact, an even larger effect on the maximum impedance can be noticed in the simulation result. The most important result of this comparison is, however,

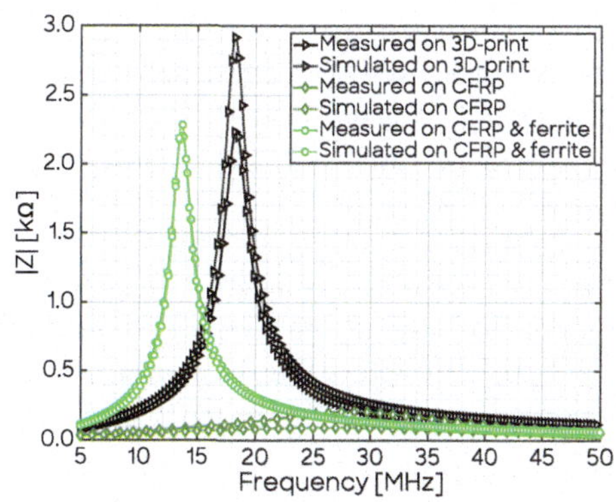

Fig. 3.24 Simulation to measurement comparison for model validation

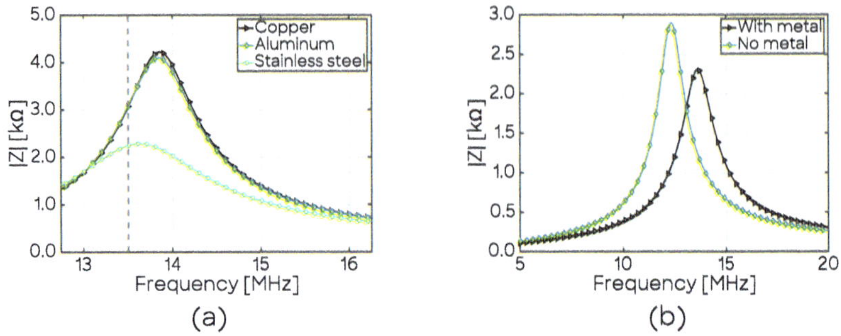

Fig. 3.25 Simulation of different configurations. **a** Comparison of different coil materials. **b** Influence of surrounding metal surfaces

that the full material stack of antenna, CFRP and ferrite results in a very good match. Therefore, the simulation model introduced here can be used to estimate the influence of different material combinations at least on the antenna resonance frequency with sufficient accuracy. The estimation of the resulting quality factor will be in the correct range but may differ slightly from the simulated results when performing real measurements. Nevertheless, this enables the model to be used for more simulations that are presented here.

As a first parameter to vary in the simulation, different conductor materials are defined for the antenna coil and the surrounding material, to investigate the influence of different conductor materials on the antenna performance (see Fig. 3.25a). It can be seen that the Q-factor is significantly higher for copper and aluminum antennas. This is to be expected because of their increased conductivity compared to stainless steel. This also means that if an FML antenna can be realized in stainless steel, other usually chosen conductor materials will only work better if chosen for the application instead, thereby verifying the material choice as a challenging one.

The second simulation is done on the change to the impedance over frequency curve when inserting the surrounding metal layers around the antenna coil. These metal areas are remains from cutting the antenna into the finished FML. They will thereby be present in the final application in any case. It can be seen that the resonance frequency is shifted to a higher value and the Q-factor decreases as soon as the conductive areas are added (see Fig. 3.25b). This is also experienced in the measurement results in Sect. 3.4.1.

The third simulation result shows the input impedance for the medium-sized antenna. When placed on a 3D-printed layer of PLA it shows a Q-factor of $Q = 6.04$ at a resonance frequency of 18.19 MHz (see Fig. 3.26a). Placing the 2 mm CFRP in between shifts the resonance frequency to 26.6 MHz and immensely attenuates the impedance. Adding the ferrite sheet between the CFRP and the antenna reshifts f_r back to 13.56 MHz. The reason for the shift can be explained by the influence of underlying materials on the inductance value of the coil (see Fig. 3.26b).

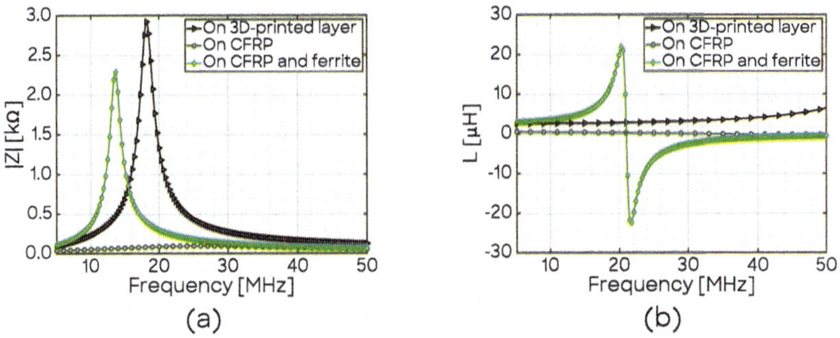

Fig. 3.26 Simulation of the antenna on three different substrates: 3D-printed clamping fixture (material PLA), directly on CFRP, and on CFRP with a ferrite sheet in between. **a** Impedance over frequency. **b** Inductance over frequency

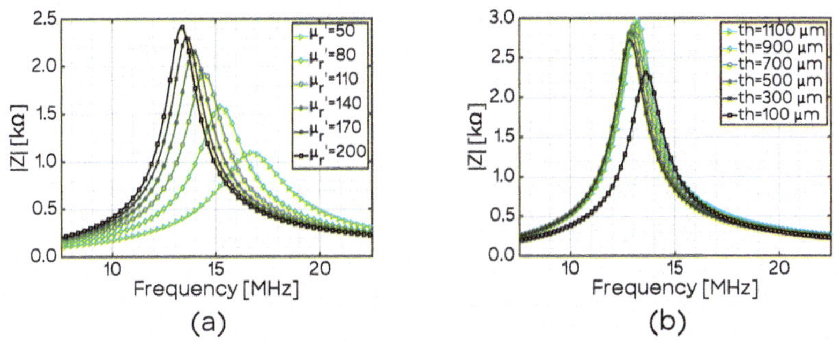

Fig. 3.27 Simulation of ferrite influence. **a** Real part relative permeability sweep, constant thickness. **b** Thickness sweep, constant permeability

The simulation allows for additional investigations of influences by slight parameter changes that cannot all be experimentally examined. The result for a 100 µm thick ferrite sheet with relative permeability swept from $\mu_r' = 50$ to $\mu_r' = 200$ in steps of $\Delta\mu_r' = 30$ (see Fig. 3.27a) shows that the higher the relative permeability of the ferrite sheet, the higher the achieved Q-factor and the lower the resonance frequency, i.e., the higher the re-tuning effect. Although the step size is chosen to be linear, the increase of the achieved quality factor is much higher for lower permeability values. Similar results can be found for a constant permeability when the ferrite layer thickness is increased from 100 µm toward 1.1 mm (see Fig. 3.27b).

References

1. S. Bornemann, J.N. Haus, M. Sinapius, B. Lüssem, A. Dietzel, W. Lang, Stainless-steel antenna on conductive substrate for an shm sensor system with high power demand. Sensors **21**(23) (2021)
2. A. Dhar, P. Pattanayak, A. Kumar, D.S. Gurjar, B. Kumar, Design of a hexagonal slot rectenna for rf energy harvesting in wi-fi / wlan applications. Int. J. RF Microwave Comput.-Aided Eng. **32**(12) (2022)
3. F. Alneyadi, M. Alkaabi, S. Alketbi, S. Hajraf, R. Ramzan, 2.4ghz wlan rf energy harvester for passive indoor sensor nodes, in *2014 IEEE International Conference on Semiconductor Electronics (ICSE2014)* (IEEE, 2014), pp. 471–474
4. B. Jiang, J.R. Smith, M. Philipose, S. Roy, K. Sundara-Rajan, A.V. Mamishev, Energy scavenging for inductively coupled passive rfid systems, in *2005 IEEE Instrumentation and Measurement Technology Conference Proceedings* (IEEE, 2005), pp. 984–989
5. D. Gould, M. Meiners, W. Benecke, W. Lang, Condensation detection using a wirelessly powered rf-temperature sensor. IEEE Trans. Veh. Technol. **58**(4), 1667–1672 (2009)
6. A. Lazaro, R. Villarino, D. Girbau, A survey of nfc sensors based on energy harvesting for iot applications. Sensors (Basel, Switzerland) Special Issue: RFID Sensor Tags: Hardware, Implementation, and Demonstrations **18**(11) (2018)
7. F. Costa, S. Genovesi, M. Borgese, A. Michel, F.A. Dicandia, G. Manara, A review of rfid sensors, the new frontier of internet of things. Sensors (Basel, Switzerland) **21**(9) (2021)
8. S. Bovelli, F. Neubauer, C. Heller, A novel antenna design for passive rfid transponders on metal surfaces, in *2006 European Microwave Conference* (IEEE, 2006), pp. 580–582
9. X. Qing, Z.N. Chen, Characteristics of a metal-backed loop antenna and its application to a high-frequency rfid smart shelf. IEEE Antennas Propag Mag **51**(2), 26–38 (2009)
10. A.-I. Petrariu, V. Popa, V.-G. Gaitan, I. Finis, Test results for hf rfid antenna system tuning in metal environment, in *Proceedings of the 13th International Carpathian Control Conference (ICCC)* (IEEE, 2012), pp. 543–546
11. I. Finis, V. Popa, A. Lavric, A mathematical approach of a hf rfid multi loop antenna for metallic environments, in *2012 International Conference and Exposition on Electrical and Power Engineering* (IEEE, 2012), pp. 679–682
12. I. Mayordomo, T. Drager, J. Bernhard, Technical challenges for the integration of passive hf rfid technology in frp composite materials, in *2011 IEEE International Conference on RFID-Technologies and Applications* (IEEE, 2011), pp. 445–450
13. M. Salas, O. Focke, A.S. Herrmann, W. Lang, Wireless power transmission for structural health monitoring of fiber-reinforced-composite materials. IEEE Sensors J. **14**(7), 2171–2176 (2014)
14. K. Kankhunthod, L. Lekawat, C. Pakasiri, Improvement of hf rfid reader performance under metallic environment using ferrite sheet, in *2017 International Symposium on Antennas and Propagation (ISAP)* (IEEE, 2017), pp. 1–2
15. J.-Y. Hsu, S.-K. Kuo, Y.-H. Hung, A study of making ferrimagnetic sheet materials for rfid applications, in *China Steel Technical Report*, vol. 23, pp. 42–45
16. M. Gebhart, R. Neubauer, M. Stark, D. Warnez, Design of 13.56 mhz smartcard stickers with ferrite for payment and authentication, in *2011 Third International Workshop on Near Field Communication*, pp. 59–64. IEEE, 022011
17. J. Donnevert, *Die Maxwell'schen Gleichungen: Vom Strömungsfeld des Gleichstroms zum Strahlungsfeld des Hertz'schen Dipols*. Springer Vieweg, 3., überarbeitete auflage ed., 2021, Lehrbuch, Wiesbaden, Germany
18. S. Paul, R. Paul, *Elektromagnetische Felder und ihre Anwendungen. Springer-Lehrbuch*, vol. 2. (Springer Vieweg, Berlin, 2012)
19. H.T. Friis, A note on a simple transmission formula. Proc. IRE (Proceedings of the IRE) Waves Electr. **34**(5), 254–256 (1946)
20. C.A. Balanis, *Modern Antenna Handbook* (Wiley, Hoboken NJ, 2008)
21. Keysight Technologies, *Using a Network and Impedance Analyzer to Evaluate 13.56 MHz RFID Tags and Readers/Writers* (2017). Application Note

22. T. Brown-Brandl, J. Maselyne, F. Adrion, A. Kapun, E.F. Hessel, W. Saeys, A. van Nuffel, E. Gallmann, Comparing three different passive rfid systems for behaviour monitoring in grow-finish pigs, in *Precision Livestock Farming '17* (2017), pp. 622–631
23. Texas Instruments Incorporated, *White paper SLYY056 - Navigating your way through the RFID jungle* (2014)
24. K. Finkenzeller, *RFID-Handbuch* (2012)
25. European Radiocommunications Committee, Sharing between inductive systems and radiocommunication systems in the band 9 - 135 khz (1997)
26. R. Paul, *Elektrotechnik 1: Grundlagenlehrbuch. Felder und einfache Stromkreise*. Springer eBook Collection Computer Science and Engineering. Springer Berlin Heidelberg, Berlin, Heidelberg and s.l., zweite, neubearbeitete und erweiterte auflage ed. (1990)
27. T. Roz, V. Fuentes, Using low power transponders and tags for rfid applications. Firmenschrift, EM Microelectronic Marin
28. S.S. Mohan, M. Del Mar Hershenson, S.P. Boyd, T.H. Lee, Simple accurate expressions for planar spiral inductances. IEEE J. Solid-State Circuits **34**(10), 1419–1424 (1999)
29. H.A. Wheeler, Simple inductance formulas for radio coils. Proc. IRE **16**(10), 1398–1400 (1928)
30. STMicroelectronics, *AN2866 Application note How to design a 13.56 MHz customized antenna for ST25 NFC / RFID Tags* (2021). Application Note
31. M.D. Bitschnau, Analysis of quality factor and resonance analysis of quality factor and resonance frequency measurements of rfid transponders: Verification and comparison of rfid measurement methods (2016)
32. R. Jacobi, E. LaCost, Antenna design guide for the trf79xxa (rev. c): Texas instruments inc. Application Report (2017)
33. 3M Company, *3M Flux Field Directional Material EM15TF Series, Technical Data* (2018)
34. J. Liu, X. Huang, Y. Ren, L.M. Wong, H. Liu, S. Wang, Galvanic corrosion protection of al-alloy in contact with carbon fibre reinforced polymer through plasma electrolytic oxidation treatment. Sci. Rep. **12**(1), 4532 (2022)
35. tesa SE - Ein Beiersdorf Unternehmen, *tesa 50600 Produkt Information* (2008)
36. https://www.atlasrfidstore.com/smartrac-minitrack-hf-rfid-wet-inlay-nxp-icode-slix. Accessed 19 July 2023
37. J.M.G. Rey, J.A. Palechor, L.F. Ariza, A.G. Rozo, F. Segura-Quijano, Coil-on-chip: Design of integrated coils for inductive telemetry system, in *2012 IEEE 3rd Latin American Symposium on Circuits and Systems (LASCAS)* (IEEE, 2012), pp. 1–4
38. S. Singha, M. Thomas, Permittivity and tan delta characteristics of epoxy nanocomposites in the frequency range of 1 mhz-1 ghz. IEEE Trans. Dielectr. Electr. Insul. **15**(1), 2–11 (2008)
39. D. Scida, M. Assarar, C. Poilâne, R. Ayad, Influence of hygrothermal ageing on the damage mechanisms of flax-fibre reinforced epoxy composite. Composit. Part B: Eng. **48**, 51–58 (2013)
40. HSM Stahl- und Metallhandel GmbH, *Werkstoffdatenblatt 1.4310 / X10CrNi18-8 Nichtrostender Stahl, austenitisch*. Accessed 28 Aug 2022
41. Ferroxcube, *Soft Ferrites Ferrite materials survey*. Accessed 29 Nov 2023
42. C. Dichtl, P. Sippel, S. Krohns, Dielectric properties of 3d printed polylactic acid. Adv. Mater. Sci. Eng. **2017**, 1–10 (2017)
43. Hexcel Corporation, *HexTow AS4* (2020). Datasheet
44. Hexcel Corporation, *Hexply 8552* (2020). Datasheet

Chapter 4
Embedded Electronics

The second part of the system developed within this work is the electronic sensor node that can be integrated into FML plates. It will be connected via the wireless interface using an antenna as described in Chap. 3 for both power and data transfer and will thereby create a link between the sensor within the material and the outside world. The main functional blocks necessary to implement this can be seen in Fig. 4.7a. The design process for the sensor node as well as an experimental validation of its suitability for the desired application will be presented within this chapter. One of the main concerns regarding the application is the effect of elevated temperature over several hours on the electronic components during FML manufacturing. An experimental evaluation on the influence when embedding electronics under these conditions will be presented here as well.

4.1 State-of-the-Art: Embedded Electronics

Embedding electronic systems into materials has been an active field of research over the last years. Designing sensor nodes to be as small as possible so they can be integrated into the material without weakening or otherwise influencing it is most often the aim. Lynch and Loh [1] provide a wide overview over different wireless sensors and sensor networks with a focus on damage detection to form material integrated SHM systems. Their results show that most of the systems developed between 1998 and 2005 are battery powered and have a height of at least 1.3 cm, most of them significantly more [1]. More recent works of Voges et al. [2] present more miniaturized sensor nodes with a remaining thickness of 1.6 mm. Although this is already thin, Lang et al. state in [3] that for function scale integration, i.e., an integration without decreasing the materials function, the integrated objects should match the surrounding material parameters in size and behavior.

S. Bornemann, *Development of a Self-Sufficient, Wireless Sensor Node Using the Outer FML Layer as an Antenna*, Mechanics and Adaptronics,
https://doi.org/10.1007/978-3-031-70648-6_4

For the integration of sensors and sensor systems into FRP, several examples can be found in the recent years [4–8]. Nevertheless, no publications on systems integrated into FMLs, which come with even higher requirements for an integration of sensors and sensor nodes, could be found.

One of the main concerns during the embedding process of sensor nodes into FML, besides surrounding conductive material, is expected to be the high manufacturing temperature in combination with applied pressure. Commercial electronic components are usually specified for operating temperatures up to 125 °C. Sporadically, components for slightly higher temperatures are available. For high temperature applications, publications can be found describing adjusted high temperature designs as well as approaches for memory protection procedures at elevated operating temperatures.

Riches and Johnston [9] state one of the main concerns for high temperature electronics as the performance degradation due to increased leakage currents. They investigate an Application-Specific Integrated Circuit (ASIC) design at temperatures of up to 250 °C with respect to changes of the interconnect materials, adhesives and changing bandgap voltage at this temperature. Au et al. [10] define extremely high temperatures for electronics to be already around 250 °C as well, while they examine a specially developed sensor connection and housing principle which is able to work actively and survive a longer storage at this temperature. In addition to this, it is able to endure surrounding pressures of over 2000 bar, much more than any component in this work will face during embedding.

Regarding the effects of elevated temperature on memory cells, Singh et al. [11] present a new flash memory approach for the use at temperatures of around 300 °C, preventing high temperature problems like storage charge leakage by design adjustments. Morgul et al. [12] state that data retention in flash memories usually suffers from elevated temperature, while data processing is better suited to be done at elevated temperatures. They recommend a split system at different temperatures for different tasks. It can be found in several works that data retention degradation over increased storage temperature can be calculated for flash memory devices using Arrhenius equation [12–15]. Therefore, this equation will be used for later experiments to estimate the influence of elevated temperature in flash memory content.

4.2 Theoretical Background: Electronic Systems

In the following sections, the basic principles used during design of the electronic sensor node will be introduced. First looking at the sensor types for the application, their amplification principles, data processing and eventually memory options and their expected degradation due to elevated temperatures.

4.2.1 Sensors

Two general sensor principles can be used for the acquisition of ultrasonic waves. The piezoelectric sensors are already quite prominent for this task. Alternatively, a piezoresistive approach using a mechanical mass element that is deformed due to wave interaction with the sensor housing can be used as well. Both principles will be introduced here briefly.

Piezoelectric Sensors

The piezoelectric effect describes the accumulation of electric charges inside piezoelectric materials due to mechanical deformation (see Fig. 4.1). The effect can be observed within non-centrosymmetric materials which have a permanent electric dipole moment [16]. Lead Zirconate Titanate (PZT) with Pb(Zr, Ti)O_3 as its empirical formula is one example for a piezoelectric material often used in scientific and commercial applications alike. It belongs to the group of poly-crystalline ferroelectric ceramics which are polarized using a strong electric field to align the randomly oriented piezoelectric crystallites within the material to gain macroscopic piezoelectric properties.

The direct piezoelectric effect can be used for a variety of sensors. For the application in this work, the GUW passing through the sensor results in a lateral bending of the material and thereby in a voltage that can be measured. The inverse piezoelectric effect, on the other hand, can be used for the actuation of Lamb waves with thin, disc shaped actuators applied to the surface. The sensor and actuator types used within the frame of this work are introduced later on in Chap. 5.

The most important characteristics of piezoelectric sensors for the layout of the sensor node are that they offer a comparably stable output over the whole bandwidth, while they do not need an extra supply voltage to work. Charges accumulate on the surface as soon as a mechanical deformation occurs. These charges thereby create a voltage difference which in contrast to other voltage differences, e.g., a laboratory voltage source, is not stable over a longer time period but decays rapidly once a load, even a high impedance one, is applied for measurement. Therefore, piezoelectric

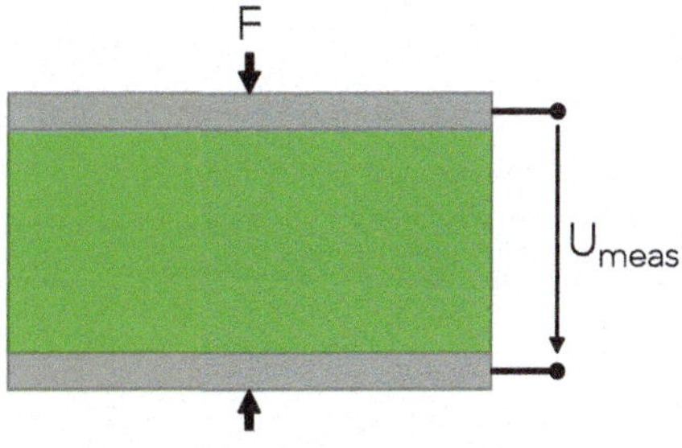

Fig. 4.1 Function principle of the piezoelectric sensor, applied force results in a charge accumulation that can be measured as a voltage difference

sensors operated at low frequencies have to be amplified using a charge amplifier instead of a regular voltage amplifier. The charge amplifier is an integrating element that gives out a voltage that depends on the charges fed into it and is therefore a charge to voltage amplifier. The working principle and advantages of this amplifier type are described in Sect. 4.2.2.

Piezoresistive Sensors

Besides the often used piezoelectric sensors described in Sect. 4.2.1, it is also possible to use the piezoresistive effect to sense ultrasonic waves inside FML. This is shown by a recently developed displacement sensor for this application [17]. The working principle is based on a seismic mass deforming a beam structure which connects it to the surrounding housing. On the deformed regions of the beam structure, there are piezoresistive doping zones which are extended or compressed, depending on the movement of the mass element. This variation leads to a difference in resistance which can be described by [18]

$$\frac{dR}{R} = 2\frac{dl}{l} - \frac{dV}{V} + \frac{d\rho}{\rho} \tag{4.1}$$

with R being the resistance, l being the length of the resistor, V being the volume and ρ being the specific resistivity of the material. Using a semiconductor based piezoresistive sensor here, the last term is the most important one. Due to deformation, the level of energy bands inside the resistor changes and thereby its specific resistance changes as well. To accomplish an intrinsic temperature compensation, as already known from many other piezoresistive sensor applications, the two piezoresistive elements are connected with two other fixed resistors on the same bulk material to form a deflection-type direct current measurement bridge (see Fig. 4.2). During Lamb wave propagation, the housing moves along with the surrounding material leading to a deformation of the inert mass. This deformation changes the resistance of the sensor elements, which can be measured as a voltage difference on the bridge voltage. With two variable resistors in the bridge, placed diagonally opposite to each other, the resulting voltage, given the variable resistance R_{var} changes in the same direction by ΔR around the base resistance R, can be calculated using [16, 19]

$$U_B = \frac{U_{Supply}}{2} \cdot \frac{\Delta R}{R}. \tag{4.2}$$

It can be seen, that the bridge voltage is directly correlated to the supply voltage of the bridge. Therefore, when designing a readout for such a system, the supply voltage should be as high as possible to get out the most amplitude of the sensor signal.

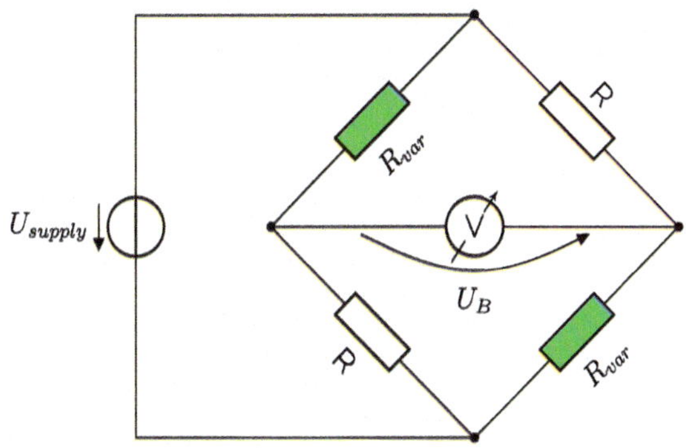

Fig. 4.2 Piezoresistive sensor with two active elements, deforming in the same way, connected in a deflection-type bridge configuration

4.2.2 *Amplifiers*

Sensor signals are usually not matching the system voltage of the digitizing unit when they are picked up at the sensor output. Therefore, for a digital sensor node it is generally not recommended to directly convert the sensor signal to a digital equivalent. If the input range is not fully used, the actual resolution obtained is worse than the theoretically provided resolution, since some bits are never influenced by the signal. This can be prevented by properly amplifying the sensor signal to match the system voltage, thereby using the optimal digitization performance. Various amplifier configurations can be used for different applications. Three amplification principles which can be used to amplify sensor signals of ultrasonic transducers are explained in the following. Two of which are later on implemented on the wireless sensor node.

Charge Amplifiers for Piezoelectric Sensors

The best way to amplify a piezoelectric sensor signal is provided by using a charge amplifier. An equivalent circuit of a charge amplifier can be seen in Fig. 4.3. An operational amplifier is the main element in this circuit. In the feedback path is a capacitor C_f with a resistor R_f in parallel. Charges that are generated by the the sensor due to the direct piezoelectric effect (see Sect. 4.2.1) are moving toward the inverting input of the amplifier and are compensated by equivalent charges (in opposite direction) from the feedback capacitor. By this, Kirchhoff's current law and the assumption that the input impedance of operational amplifiers is infinitely high are both fulfilled. This results in the following output voltage [20, 21]

$$v_{out} = -\frac{q_i}{C_f}. \tag{4.3}$$

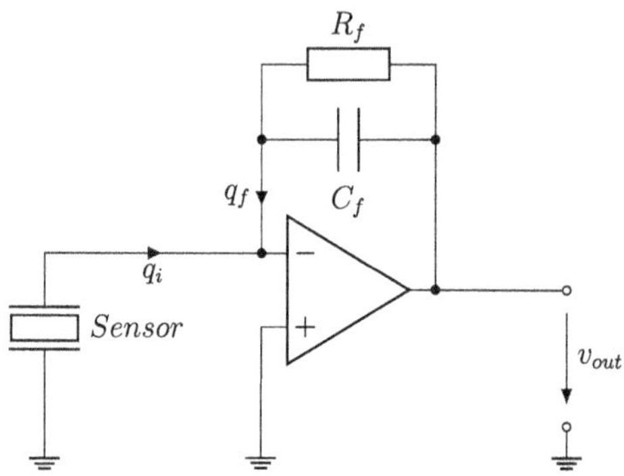

Fig. 4.3 Circuit diagram of a charge amplifier (own representation after [21])

C_f is then discharged by R_f at the limiting lower frequency of operation with

$$f_g = \frac{1}{2\pi R_f C_f}. \tag{4.4}$$

The clear advantage of a charge amplifier is given in Eq. 4.3. It specifies that the output voltage only depends on the feedback capacity, while other capacitive effects, e.g., from cables are completely irrelevant for the measurement.

Instrumentation Amplifiers for Bridge Voltages

The signal amplitude when measuring piezoresistive sensors can be comparably small. Given Eq. 4.2, especially for low supply voltages, an amplifier is required to increase the bridge voltage sufficiently to be processed further. For this purpose, a simple differential amplifier based on an operational amplifier could theoretically be used. This circuit however has a comparably small input impedance compared to other amplifier configurations and is therefore not best suited for this measurement purpose. Additional input buffers using either voltage followers or a combination of two non-inverting amplifiers can help with this problem. The latter solution is commercially available as a monolithic component and is called instrumentation amplifier.

For deflection-type bridge measurement applications, instrumentation amplifiers are by far the best choice. They are designed to selectively amplify the differential voltage between both inputs, while rejecting common mode offsets. An instrumentation amplifier consists of a buffer stage followed by a differential amplifier (see Fig. 4.4). The design is often based on resistor pairs, which are marked here as $R_2 = R_2'$, $R_3 = R_3'$ and $R_4 = R_4'$. With this, it is possible to adjust the gain of the instrumentation amplifier by changing one single resistor R_1.

Both non-inverting amplifiers operate as voltage buffers for differential signals, providing high input resistance thereby not influencing the sensor output. Common mode parts to both inputs are only passed through with unity gain because both

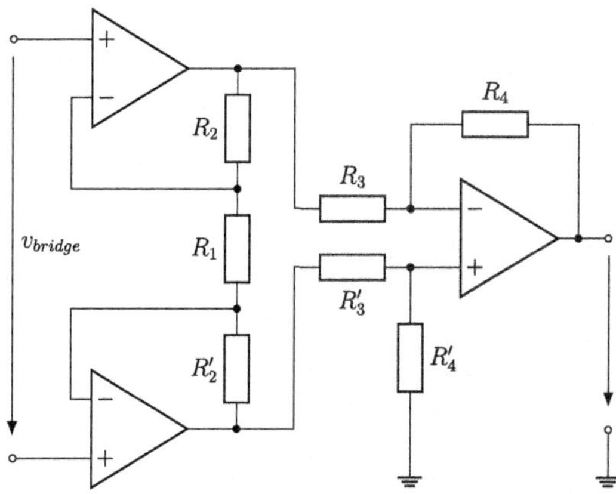

Fig. 4.4 Instrumentation amplifier circuit: Left side buffer stage, right side differential amplifier stage

outputs result in the same offset voltage. The current through resistors R_1 and both R_2 is equal, resulting in an equal voltage drop across both R_2 resistors. Thereby, the resulting voltage drop across R_1 equals the voltage difference at the inputs. In case both inputs have the same potential, there will be no voltage difference over R_1, the output voltage will be 0 V. Given the differential character of the voltage across R_1, the gain of the circuit can be adjusted by changing the resistance value of R_1.

The output of the differential amplifier is the difference between its inputs, which is at the same time the difference between the input stage's outputs and will be amplified by the ratio of R_3 and R_4. This gives the output voltage of the complete instrumentation amplifier as

$$v_{out} = -v_{bridge} \cdot \left(1 + \frac{2R_2}{R1}\right) \cdot \left(\frac{R_4}{R_3}\right) \tag{4.5}$$

with R_2, R_3 and R_4 being constant in most cases, resulting in a total amplification of

$$G_{inamp} = \frac{v_{out}}{-v_{bridge}} = \left(1 + \frac{2R_2}{R_1}\right) \cdot \left(\frac{R_4}{R_3}\right). \tag{4.6}$$

Some instrumentation amplifiers offer the possibility to apply a reference voltage so the output at $v_{bridge} = 0\,\text{V}$ is the reference voltage, providing easier data handling for a connected Analog-to-Digital Converter (ADC).

Non-inverting Amplifier

The non-inverting amplifier is another type of amplifier that is frequently used to increase the signal amplitude of sensor signals. It is simpler than the instrumentation amplifier, but well suited, e.g., to amplify the signal that comes from a previous amplifier stage. The circuit diagram of a non-inverting amplifier (see Fig. 4.5) shows

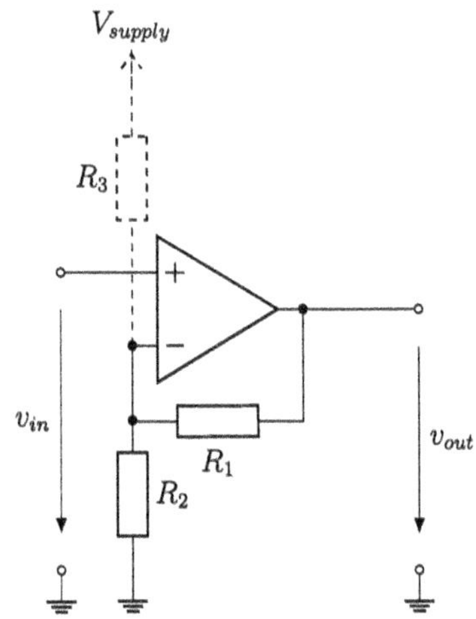

Fig. 4.5 Non-inverting amplifier circuit with input offset correction

the feedback path consisting of a voltage divider of R_1 and R_2 which provides part of the output voltage to the negative input of the amplifier. According to the assumptions that can be made for operational amplifiers, both input nodes have the same electrical potential. The amplification of a non-inverting amplifier can therefore be calculated by

$$G = 1 + \frac{R_1}{R_2}. \tag{4.7}$$

Thereby, the output voltage v_{out} can be calculated by

$$v_{out} = G \cdot v_{in}. \tag{4.8}$$

Due to the working principle described above, a constant voltage offset that is present at v_{in} would be amplified with the signal as well. This offset is not filtered out by the common mode rejection capabilities of the amplifier and must therefore be compensated in another way. This scenario occurs, e.g., when the non-inverting amplifier is put as a second amplifier stage after an instrumentation amplifier having an output voltage around a given reference voltage. A simple solution to this is to apply the constant offset to the feedback path. This offset compensation can be done for a known input offset voltage v_{oc} using an additional resistor R_3 and the supply voltage (see Fig. 4.5, dashed part). With this, the offset voltage applied at the inverting input of the amplifier can be calculated using Thevenin's Theorem to

$$v_{oc} = V_{supply} \frac{R_2}{R_3 + R_2}. \tag{4.9}$$

Thereby, the gain is modified to

$$G_{oc} = 1 + \frac{R_1}{\frac{R_2 \cdot R_3}{R_2 + R_3}} = 1 + \frac{R_1 R_2 + R_1 R_3}{R_2 R_3}. \tag{4.10}$$

This means, when using a voltage divider with $R_2 = R_3$, the gain of the circuit almost doubles compared to Eq. 4.7 and the output voltage is set around v_{oc}.

Gain Bandwidth Product

Besides the low power and high gain feature, one of the most important characteristics of amplifiers used in this work will be the Gain-Bandwidth Product (GBWP). It is also known as the unity gain bandwidth, i.e., the bandwidth for which the amplification is one [16]. This means, if a device has a GBWP of 30 MHz, when it is configured for unity gain it will work as configured at frequencies up to 30 MHz. Is it going to be used as an amplifier with an assumed gain of 100, it will only show this gain to a frequency of 300 kHz. At higher frequencies, the implemented gain will reduce regardless of the configuration circuitry.

This correlation can be expressed by

$$G_{max,BW} = \frac{GBWP}{f_{max}}, \tag{4.11}$$

stating that for a lower maximum operation frequency, i.e., a lower bandwidth of the amplifier, a higher gain can be configured.

4.2.3 *Analog-to-Digital Conversion*

Analog-to-digital conversion is needed any time an analog signal is intended to be processed on a digital processing unit like a microcontroller. There are several methods to digitize signals. Mainly successive approximation conversion and delta-sigma conversion are used for ADC architectures in microcontrollers nowadays. The approach used in the developed sensor node, successive approximation data conversion, will be explained here.

Successive Approximation Data Conversion

The successive approximation approach for analog to digital data conversion (see Fig. 4.6) uses a sample and hold circuit (S&H) to hold one constant input voltage (v_{in}) value from start to end of the conversion [19]. This voltage is then fed into a comparator (Comp), which compares this voltage to successively applied reference

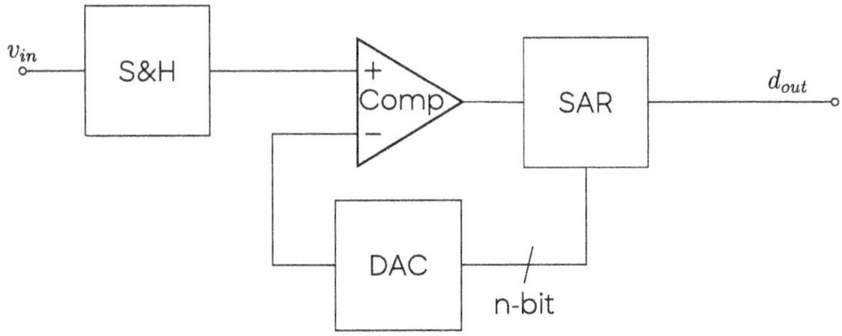

Fig. 4.6 Schematic diagram of the SAR ADC working principle. With block "SAR" being the successive approximation register, "S&H" being the sample and hold circuit and "DAC" being the digital to analogue peripheral for reference voltage generation at the comparator

voltages (v_{ref}), giving a binary information if the comparison resulted in $v_{ref} < v_{in}$ or the opposite. The reference voltage is provided by the Successive Approximation Register (SAR), feeding a bit combination to a Digital-to-Analog Converter (DAC). The start of conversion always begins with setting the Most Significant Bit (MSB) to a *HIGH* logic level, thereby generating half of the supply voltage for the first comparison $v_{ref} = v_{supply}/2$. If the comparator gives out the MSB set results in $v_{ref} < v_{in}$, the MSB stays *HIGH*, otherwise it is set to *LOW* and the second bit is set to *HIGH*. This continues for all bits of the SAR. By using bigger reference voltage steps at the beginning and smaller steps later on, this method is very fast for all given ADC resolutions.

Sampling Rate and Nyquist–Shannon Sampling Theorem

Most ADCs require some time in which the sample and hold circuit is active and the signal is converted. This is known as the sampling interval t_s. The maximum achievable sampling frequency f_s is given as [19]

$$f_s = \frac{1}{t_s}. \tag{4.12}$$

For most applications, converting analog to digital signals is a trade-off between storage capacity, time of conversion, resolution, and sampling rate. For the sampling rate, there is an orientation provided by the Nyquist–Shannon sampling theorem. It states that for a band limited signal with a maximum frequency of f_{max}, this signal can be reconstructed from a sequence of equidistant points sampled with [19]

$$f_s > 2 \cdot f_{max}. \tag{4.13}$$

This is a very important correlation for the design choice of the sampling rate. The maximum signal frequency which determines the minimum required sampling rate for signal reconstruction and allows to theoretically chose a maximum required sampling rate to lower the power consumption.

4.2.4 *Voltage Regulation for Embedded Sensor Nodes*

For voltage regulation, there are several approaches available. The classical AC/DC and DC/DC switching converters like Buck or Boost converters have the advantage of higher power efficiency compared to linear voltage regulators. They switch the input voltage over a combination of reactive components and diodes to generate the targeted output voltage. The main drawback with this is a high output voltage ripple resulting from the switching principle. This ripple is usually significantly bigger than for linear types. Linear regulators on the other hand provide the advantage of a smoother output voltage with the disadvantages that the voltage difference $v_{in} - v_{out}$ is completely dissipated as heat, which leads to a decreasing efficiency the higher this difference is. In addition, due to their working principle, they are not able to convert a lower input voltage to a higher value. Nevertheless, for the application in this work, a linear regulator is a far better choice than a switching one.

Low Drop Out Regulators

One of the most often used linear regulators in combination with microprocessors is the Low Dropout Regulator (LDO). This regulator requires only a small voltage drop, i.e., a small voltage difference to work. Hence, for low power systems, they are mostly advantageous. The efficiency for those systems is even quite sufficient because the lower the voltage difference, the higher the efficiency for a system converting excess energy to heat. AnLDO passes the input voltage over a p-type transistor in its linear region, where it acts as an active load. Its channel resistance is controlled by an error amplifier which increases control voltage or current, depending on the transistor type, according to a provided reference voltage.

4.2.5 *Effects of Elevated Temperature on Sensor Nodes*

One of the main concerns for the material integration of the sensor nodes into FML is the elevated temperature of 180 °C over several hours and its influence on the electronics function. The direct impact will be investigated experimentally in a later section, while the theoretical estimation of elevated temperatures on the system lifetime, focusing on memory lifetime, will be done based on theory.

The JEDEC Standard JESD91B [13] lists the Arrhenius equation

$$A_T = e^{(\frac{-E_{aa}}{k_B})(\frac{1}{T_t} - \frac{1}{T_s})} \tag{4.14}$$

as the one applicable to describe the correlation of temperature and failure in most semiconductor failure mechanisms. Therefore, for electronic components, it describes the accelerated aging due to increased temperature. With A_T being the acceleration factor due to changes in temperature, E_{aa} being the apparent activation energy, the Boltzmann's constant k_B and T_t and T_s as the absolute temperature of the experiment and of the system lifetime reference, respectively.

4.3 Sensor Node Design

In this section, the development process for the sensor node is presented. Every component (see Fig. 4.7) is introduced and features that led to the selection are highlighted. The main idea is to use commercially available components for system assembly, to ensure easy and reliable reproduction. Also, if the system can be setup to work in the desired way with off the shelf components, a specialized ASIC is expected to perform even better. The targeted overall height is set to be smaller than 1 mm, with no limitation at the lower end. The development steps over different PCB configurations are presented briefly. The final sensor node prepared for embedding has an outer diameter of 17 mm × 17 mm and a height of less than 800 μm.

4.3.1 *Sensor Node Versions*

The first version of the sensor node (see Fig. 4.8a) is designed for a general functional test on a 1.55 mm rigid Printed Circuit Board (PCB). In this version, external ADC

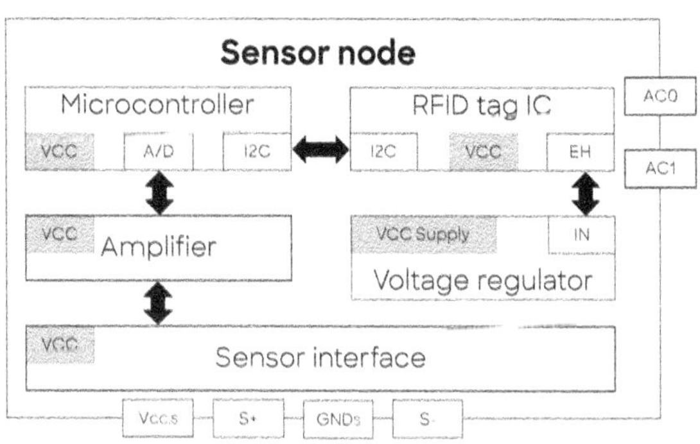

Fig. 4.7 Functional block diagram of the wireless sensor node for material integration

(a) (b)

Fig. 4.8 First sensor node versions. **a** Sensor node 1, designed for testing of function and capabilities of first component choice. **b** Sensor node 2 on a round, flexible PCB

ICs are validated, which are able to sample with maximum rates of 3 mega samples per second. The amplifier stage is missing in this version. There are several connectors on it for part-wise functional unit tests and later on connections using bridging jumpers. Lateral dimensions are 35 mm by 26 mm with a height, excluding connectors, of 3 mm. These dimensions are not suited to be embedded into FML plates, which are intended to be 2–4 mm thick in total, but for evaluation purposes this is fine.

The results of this evaluation show that the general setup is working and that microcontroller and RFID IC are well suited for the task. The voltage regulator IC will be changed for a different one due to better package handling for manual assembly. The external ADC ICs however are not used in the following designs, because the additional power consumption caused by them is too high. They will be substituted by the microcontroller's internal low power compatible ADC instead.

After the first version, a second version with a reduced amount of components is designed and manufactured on a flexible PCB (see Fig. 4.8b). It is a round design, according to previous designs for material integration found in literature. The general idea behind a round design is to reduce mechanical influences on the surrounding material similar to the reasoning of a round antenna shape found in Chap. 3. This PCB with a diameter of 23 mm is made from adhesively bound polyimide layers. This adhesive can withstand permanent temperatures of up to 120 °C and is thereby only suited for embedding experiments into GFRP, as this material combination is cured at lower temperatures. The first embeddings are performed using this sensor node version. The previously present pinheader for programming is exchanged by eight spring contact pads instead. In addition to that, the antenna connection is done using an miniature (U.FL/UMCC) connector for coaxial cable attachment. The third version of the sensor node (see Fig. 4.9a) is made from most components listed in detail in the following sections. It only lacks a second amplifier stage, which is necessary to sufficiently amplify also the smaller piezoresistive sensor signals. Therefore, this version will mainly be used for measurements with the piezoelectric sensors. It is manufactured on a temperature resistant, adhesiveless, flexible PCB. This allows for material integration at 180 °C without any adhesion problems on the substrate side. It can be seen, that it has holes distributed over its edge area to allow resin to flow through, so that a better connection to the surrounding material can be achieved.

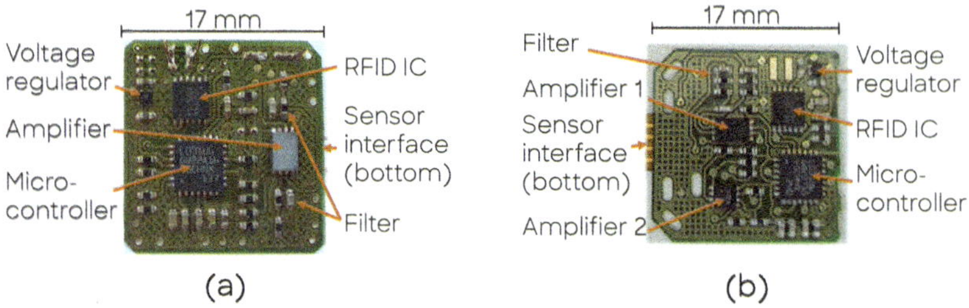

Fig. 4.9 The last sensor node versions. **a** Labeled sensor node V3. **b** Labeled sensor node V4

The geometrical design is done as a rectangular shape again for several reasons: The first one is that cutting prepreg layers in shape of the sensor node to allow for a plane stack-up in the end is much harder to achieve for a round shape than for a rectangular one. The second reason is given by the combination of available component packages and the thereby best achievable area reduction by placing them closely together in a square or rectangular shape. With a round approach there are always areas which are provided for the sake of being round but not used because of space requirements by the components or interconnections. The third reason is that, regarding an integration into the FRP layers and the anisotropy of the material due to fiber direction, all fibers cut in a line are weakening the material by far more than any right angled resin formation around edges ever could. Therefore, reducing area in total is much more interesting at this point than having a round shape. Nevertheless, the corners of the PCB are still designed rounded to avoid sharp 90° angles.

For the programming connection, again new approach is chosen for this and all following versions: a soldered header that can be attached to one side of the PCB during assembly and testing. Once the sensor node is ready to be embedded, it is simply cut off and the PCB is in its final shape of 17 mm by 17 mm with a total height of less then 800 µm. The results obtained with this version can be seen for the piezoelectric measurements in Sect. 5.4.2, as well as some of the preliminary piezoresistive measurements close to the sensor's resonance frequency shown in Sect. 5.4.3 as well.

The final version adds a second amplifier after the one already present on PCB version 3, for proper amplification of the piezoresistive sensor signals. The general PCB shape is the same as for version 3, as it has proven good assembly and handling capabilities. The components are placed closer together in this version, allowing for bigger holes for resin flow in the region around the sensor connection (see Fig. 4.9b). The results obtained with this version for are shown in Sect. 5.4.4.

4.3.2 RFID Tag IC

The first and at the same time one of the most important system ICs is the RFID tag IC. It works as the interface between antenna and sensor node. Based on the results presented in Sect. 3.2.3, the sensor node is going to be wirelessly connected using HF-RFID. With this, a small amount of power as well as data can be wirelessly transmitted and an additional power supply, as known from many mobile systems, will not be required.

The functional components of the 550 µm thick RFID-IC *ST25DV64K* can be seen in Fig. 4.10. In addition to the analog front-end and internal Electrically Erasable Programmable Read-Only Memory (EEPROM) memory, the circuit has an internal voltage regulation to 1.8 V to supply its internal subsystems. This voltage regulator does however not affect the energy harvesting output. The minimum operating field strength for simple data readout is in accordance to ISO standard 150 mA/m, the maximum recommended field strength is 5000 mA/m. While the antenna is connected to the AC0 and AC1 pin of the IC, the energy harvesting unit of the analog frontend delivers a voltage to the V_{EH} pin. This voltage is then fed into the next component, the voltage regulator, described in Sect. 4.3.3. To use the digital unit control, which hosts also the communication bus Inter-Integrated Circuit (I2C), an external voltage (V_{CC} to V_{SS}) of at least 1.8 V has to be supplied that is taken from the output of the voltage regulator. The SCL and SDA pins are clock and data pins for the I2C and are connected to the supply voltage via a 2 kΩ pull-up resistor each. The communication bus can be operated at data transmission rates of up to 1 MHz to enable fast sensor data transfer. The analog frontend contains the capacitive part of the LC-resonance circuit that is required for HF-RFID applications, having a so called tuning capacitance of around 28.5 pF. This capacitance is taken during antenna design for first geometry calculations in Sect. 3.3. For communication between the microcontroller and the RFID IC, the I2C bus is used at its maximum frequency of 1 MHz.

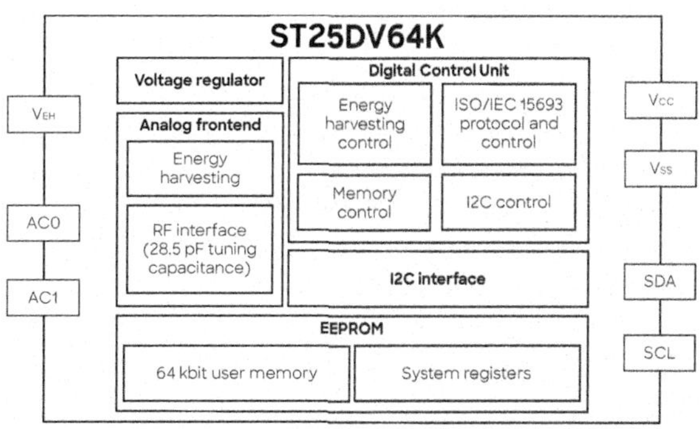

Fig. 4.10 Block-diagram of important internal components of the RFID IC. Own representation according to [22]

Energy Harvesting

The energy harvesting mode is turned on and set to active by default once initially for every IC in use. After that, the energy harvesting output delivers power according to the applied magnetic field close to the antenna. The results of investigations on energy harvesting capabilities in combination with different antennas are presented in Sect. 3.6.

Activating and loading the energy harvesting output can lead to a prevention of memory readout during the time energy is harvested. The manufacturer provides additional material to the IC, stating that if the power consumption on V_{EH} is high, the tag IC stops to communicate with the reader. Also, the amplitude modulation ratio has an influence on the maximum consumed power during active communication. When 10% amplitude modulation is used at maximum field strength, up to 12.59 mW of power can be actively supplied to other systems by V_{EH} without interrupting communication capabilities, while with 100% amplitude modulation the maximum is given with 8.64 mW. The experiments have shown that it is in fact not possible to read out the memory of the RFID tag IC while the measurements or memory write action is performed by the sensor node. This capability recovers immediately, once the program is finished and the current consumption reduces to a lower value. The results of power consumption measurements can be seen in Sect. 4.5.2.

Data Transmission Approach

For wireless data transmission to and from the RFID-IC, generally two possibilities open up. One is the fast transfer mode, it provides 256 byte of memory buffer that can be used for a very fast data transfer of short information at 26.5 kb/s for writing via the wireless link and up to 53 kb/s for wireless reading. This option is not used within this work but could be later on used to send, e.g., a trigger signal toward the material integrated sensor node.

The data transmission approach chosen for this work is using the RFID IC's internal 64 kbit EEPROM. It offers a data retention of 40 years and 1 million write cycles at 25 °C. Regarding the intended application, where the sensor node will be active every once in a while during maintenance, saving the measurement values to the EEPROM and reading it out after the measurement cycle is chosen as the best method available. To reach the amount of write cycles during regular inspection intervals is very unlikely, even if it is assumed that each sensor node is read out for damage detection multiple times during the same maintenance cycle.

4.3.3 Voltage Regulation

The voltage regulator chosen for the sensor node is the TPS7A03 with up to 6 V of input voltage and a constant output voltage of 1.8 V at up to 200 mA. It is a low drop out regulator, which means it only needs a small difference between input and output voltage to work properly. The working principle of this type of regulator can be found in Sect. 4.2.4. In case of the TPS7A03, dropout voltage is only 270 mV at maximum load, instead of more than 1.5 V typical for other voltage regulator types. In addition to that, LDOs offer a very clean output voltage with very low noise, as they are a subclass of linear regulators (see Sect. 4.2.4). The operating temperature of the device is specified between −40 and 125 °C storage temperature is given with up to 150 °C. The chosen package is the DQN-X2SON with a footprint of 1 mm × 1 mm and a maximum package height of only 400 µm.

The combination of small space requirements and its low dropout voltage make this regulator the perfect fit for the application on the sensor node. With this, the harvested voltage is controlled to stable 1.8 V, making the system independent from applied RFID field strength at the antenna which would otherwise result in a load-dependent voltage. In addition to this, the theoretical low noise level of the output makes it perfectly suited to supply the measurement bridge of the piezoresistive sensor approach, as well as the reference voltages used by the amplifiers and the ADC. The only additional elements that are needed are two capacitors, one from input to ground for good transient response and input ripple suppression and one at the output mainly for stability reasons as already explained in Sect. 4.2.4. Although theoretically a low noise signal is provided at the output, measurements show that this is not the case for a harvested Radio-Frequency (RF) voltage as displayed in Fig. 4.11. For a laboratory voltage supply, a peak-to-peak voltage ripple of 30 mV around 1.82 V can be measured while for RF energy harvesting this peak-to-peak voltage ripple increases to 158 mV around 1.81 V. It can be assumed that this higher ripple occurs due to a high-frequency ripple also present on the input voltage. For the input voltage

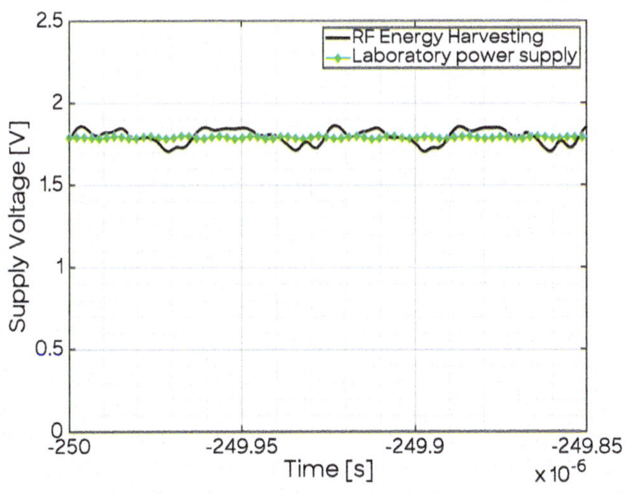

Fig. 4.11 Output voltage at the low dropout regulator for laboratory voltage source and RF harvested energy compared

the peak-to-peak ripple can be determined as 281 mV around a harvested voltage of 2.49 V in an exemplary measurement. It is to be expected that there is more noise coupled into the circuit when an RF field is present close by. This even increases if RF energy is harvested from the field and fed into the regulator to supply the sensor system. This can be managed, but it has to be kept in mind when it comes to measurements done with the circuit as filters for the sensor signal will be required.

4.3.4 Microcontroller

The microcontroller that controls the measurement procedure of the sensor node is the STM32L031G6U6. It is an ultra-low-power device working at down to 1.65 V supply voltage and can be set for an operation at 76 μA MHz^{-1}. Given the low power requirement set up by the decision to use RF energy harvesting as power source, this microcontroller is a good choice to control the measurement task. In addition to this, the internal RC clock source offers a system clock frequency of up to 16 MHz, which is sufficient to digitize sensor signals at a high rate and to perform on-chip calculations on the digitized signals.

The feasibility of attaching an external ADC which provides an even higher sampling rate has been investigated but comes with the price of a significantly increased power consumption. This is caused not only by the external component itself but also by the data transmission toward the microcontroller. Therefore, the external ADC approach is not chosen for this sensor node.

To perform the analog to digital conversion, a 12-bit ADC is available inside the microcontroller and is used instead of an external one. It comes with a successive approximation ADC and is able to sample with up to 12 bit resolution and up to 1.14 Mega samples per second (Msps) at this resolution. With this it is meeting the requirements of a sufficient sample rate for the application. To keep the data processing as simple as possible, the resolution is scaled down to only 8 bit which is sufficient as long as the sensor signal is properly amplified. Data storage on the RFID IC is done in word wise manner, therefore enabling the data to be directly transferred if already present in 8 bit resolution. In addition to this, the ADC conversion timing decreases significantly. The total conversion time per value can be calculated using equation

$$t_{conv} = (c_s + c_c) \cdot \frac{1}{f_{ADC}} = (3.5 + 9.5) \cdot \frac{1}{16\,\text{MHz}}. \tag{4.15}$$

For an 8-bit resolution the total conversion time is 0.8125 μs. The required conversion cycles c_c depend on the chosen resolution while the minimum sampling cycles needed c_s are determined by the output impedance of the analog signal source. On the selected microcontroller there are multiple ADC input pins defined, three of them with the "fast channel" property: PA0, PA4 and PA5. This property describes the ability of those inputs to allow high-speed conversions even for analog signal supplies with higher output impedance, i.e., higher ADC input impedance. The selected

Table 4.1 Maximum input impedance for ADC peripheral clock frequency of 16 MHz based on [23]

T_s (cycles)	t_s (μs)	R_{AIN} max fast channels (kΩ)	R_{AIN} max for standard channels (kΩ)			
			$V_{DD} > 2.7$ V	$V_{DD} > 1.8$ V	$V_{DD} > 1.75$ V	$V_{DD} > 1.65$ V, $T_A > 25\,°C$
1.5	0.09	0.5	<0.1	NA	NA	NA
3.5	0.22	1	0.2	NA	NA	NA
7.5	0.47	2.5	1.7	NA	NA	NA
12.5	0.78	4	3.2	NA	NA	NA
19.5	1.22	6.5	5.7	NA	NA	<0.1
39.5	2.47	13	12.2	NA	NA	5
79.5	4.97	27	26.2	<0.1	NA	19
160.5	10.03	50	49.2	32	<0.1	42

instrumentation amplifier has a typical output impedance of 0.6 kΩ. An overview of maximum allowed ADC input impedance (R_{AIN} max) for fast and standard channels is given in Table 4.1. Using a fast channel, a minimum sampling time of 3.5 cycles can be realized leading to a sampling frequency of

$$f_{sample} = \frac{1}{t_{conv}} = 1.23 \text{ Msps}. \tag{4.16}$$

Given the Nyquist–Shannon sampling theorem by Eq. 4.13, with this sampling frequency it is theoretically possible to sample signals up to 615 kHz. For the application within this work signal center frequencies of up to 300 kHz are expected to be used, hence this sampling rate will be sufficient to reconstruct the measured Lamb waves from the digital data. The ADC values provided by the ADC peripheral are then directly forwarded to memory using the so called Direct Memory Access (DMA). This allows to quickly shift data between two peripherals without the need of a processor action. With this, the processor can be used for other tasks or simply powered down to reduce energy consumption, while data is transferred in the background.

Another task that is done by the microcontroller is the bus communication toward the RFID IC with which the data is written onto the internal EEPROM for later wireless readout. For this, the I2C bus is used at maximum possible speed setting. After the data conversion is completed, the data is completely transferred to the EEPROM and after that, the microcontroller is put into stop mode, where it consumes less energy. This is necessary because during the active time of the sensor node no wireless memory readout is possible due to the increased amount of power that is required from the energy harvesting output for supply.

In addition to the analog-to-digital conversion and data communication tasks, the microcontroller also activates the piezoresistive sensor by pulling the power supply General Purpose Input Output (GPIO) to a *HIGH* state. This can be used to supply the sensor's deflection-type measurement bridge, right before it is measured and deactivate it again after the measurement is done. With this, it is possible to reduce the amount of energy that is required by the sensor node whenever no measurement task is performed. For assemblies that are prepared for piezoelectric sensors, this functionality can be deactivated because the sensor principle works without any additional supply voltage.

4.3.5 Signal Conditioning

The sensor signal, regardless of the type of attached sensor, is small compared to the system voltage. This means for proper digitization it has to be amplified first, especially if a low resolution of 8 bit is used. In addition to this, there are some noise sources around the plate which can have an impact on the sensor signal on its way to the sensor node. To get rid of them to the best possible amount, before they are amplified together with the sensor signal, a filter stage containing a differential band-pass filter is applied right after the sensor interface. The filter as well as the amplifier stage used are introduced in the following.

Filter

Measuring during the application of a nearby magnetic field with a frequency of 13.56 MHz can lead to increased noise on the measurement. However, this noise will have a higher frequency than the expected sensor signal. A possible Direct Current (DC) offset in the measured voltage should already be suppressed by the intrinsic common mode rejection of the amplifier, but for very small differential signals, filtering out a good amount before still improves the performance. Therefore, a bandpass filter is inserted between the sensor and the amplifier to keep any unnecessary signal noise from being amplified. The lower cut-off frequency of the high pass is set with a combination of a 15 nF capacitor and a 10 kΩ resistor resulting in

$$f_g = \frac{1}{2\pi RC} = 1\ \text{kHz}. \tag{4.17}$$

Following the high pass filter, a differential low pass filter built from 10 kΩ and 15 pF with a cut-off frequency of

$$f_g = \frac{1}{2\pi(10\ \text{k}\Omega + 10\ \text{k}\Omega) \cdot 15\ \text{pF}} = 531\ \text{kHz} \tag{4.18}$$

is used to suppress the RFID noise.

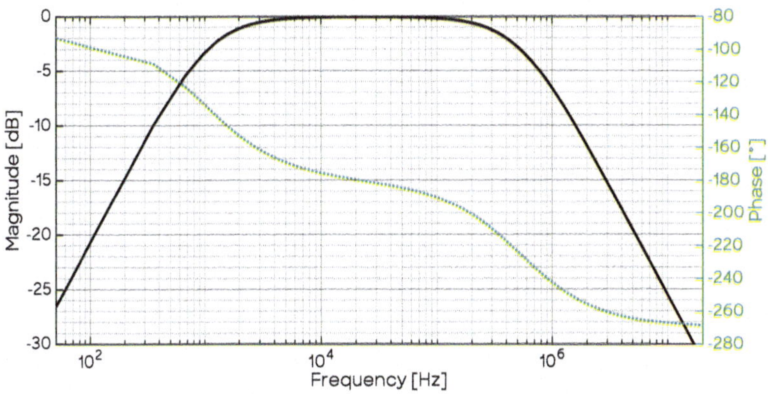

Fig. 4.12 Bode plot of the filter stage between sensor and amplifier to eliminate low-frequency offsets as well as high-frequency RF noise

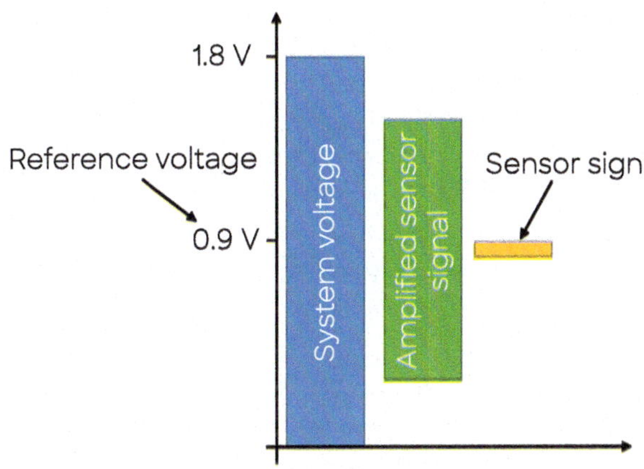

Fig. 4.13 Schematic comparison of the sensor, system voltage and targeted amplified signal amplitude to provide a sufficient resolution using 8-bit

For verification of the calculations, the filter stage is simulated using an AC analysis in LT-Spice, which confirms the calculated frequency band (see Fig. 4.12).

Amplifier Stage

One of the amplifiers used in this project is an instrumentation amplifier, the *MCP6N11* from *Microchip Technology* in the 100 V/V subtype [24]. As already introduced in Sect. 4.2.2, it is suited for amplifying bridge voltages and piezoelectric signals alike. It works at a low supply voltage of 1.8 V and has a specified GBWP of 35 MHz, a common mode rejection ratio of more than 88 dB. This means, at the highest Lamb wave frequency targeted to be used within the scope of this project of 300 kHz, it still provides an amplification of up to G = 116 following Eq. 4.11. The main drawback with this IC is its packaging. It is specified to have a maximum height of 800 µm, which is 250 µm higher than all other components. Therefore, a height reduction for this package will be investigated in Sect. 4.5.1. For piezoelectric sensors, measurements show that this single amplifier is fully sufficient (see Chap. 5). For the piezoresistive measurement however, given a low bridge supply voltage, the signal amplitude above the sensor's first resonance frequency is significantly smaller than that of the piezoelectric sensors (see Fig. 4.13). Therefore, a second amplifier stage is cascaded after the first one so that a final amplification of

$$G_{total} = G_{INA} \cdot G_{OPAMP} \tag{4.19}$$

can be achieved. The drawback comes with an additional amplification of the noise output from the first amplifier and a higher overall power consumption due to another IC.

As second amplifier, the TSV7721 [25] is chosen with a GBWP of 22 MHz, and a required supply voltage of 1.8 V. It comes in an even smaller package than the instrumentation amplifier so no modifications are required here.

Figure 4.13 shows the importance of amplification for the application. The system voltage is chosen to be 1.8 V which defines also the maximum for the ADC peripheral. For any fixed resolution of N bit, the smallest distinguishable voltage difference is

$$\Delta V = \frac{V_{max}}{2^N} \tag{4.20}$$

where V_{max} is the supply voltage. If a low resolution of only 8 bit is chosen due to measurement time and memory limitations, the smallest ΔV is 7 mV. In case the sensor delivers only small output voltages below this value, the signal is not digitized at all. For voltages above this but still way smaller than the system voltage, only part of the available range would be used. Some bits would never change their value. Therefore, it is desirable to amplify the sensor signal to a level where it is in the range of the system voltage to use the given resolution to the best amount.

4.3.6 Sensor Interface

As the title of this work already announces, a sensor node is presented which also includes an element to fulfill the purpose of sensing Lamb waves for damage detection. The developed system is capable of using two completely different sensor types for this, with only an adjusted amplification factor. In theory, it is also possible to connect and read out even more sensor types, like thermoresistive and thermoelectric ones, but the other requirements for design, mainly the high sampling rate, are usually not required for thermal sensors. Measurements are done successfully with both sensor types as shown in Chap. 5.

The main difference in the sensor node design to be able to work with both sensor types is that the piezoresistive approach works using a measurement bridge, as already introduced in Sect. 4.2. This means the sensor node has to supply the bridge to be able to measure the resulting bridge voltage. A purely analog approach as shown in [26] is not suitable for this. Therefore, a digital sensor node is designed containing a microcontroller as a controlling element.

The piezoelectric sensors used in this work are *DuraAct P-876.SP1* from manufacturer *PI Ceramic* with lateral dimensions of 16 mm times 13 mm and a thickness of 500 μm, of which 200 μm are the functional part with a capacity of 8 nF.

Sensor Connection

While the piezoelectric sensor is equipped with two solder pads which are 5 mm apart and can be easily connected to the sensor node using enameled copper wires, for the resistive sensor type a connection PCB has to be designed. The first attempt (see Fig. 4.14a) is manufactured at the IMSAS cleanroom. Assuming at first, that an inlay made from polyimide could influence the wave propagation inside FML, the aim is to achieve a very thin result to keep that influence as small as possible. For metallization, a layer of gold is sputtered onto the first polyimide layer as conductor and pad material. The pad metal thickness is galvanically increased to 3 μm. A second polyimide layer acts as an electrically insulating cover layer, while the pad areas are etched free for contacting afterwards. The resulting connection PCB is 17.1 mm long, 4.2 mm wide and has a thickness of 10 μm. A very good mechanical and electrical connection can be achieved using low-temperature soldering paste melting at 140 °C (see Fig. 4.15a).

Since the aim is to embed the whole system into FML at 180 °C, this low-temperature soldering paste is not suited for the application. Therefore, other products with significantly higher soldering temperatures are tested as well. The results show a loss of electrical and a decrease in mechanical connection for almost all soldered specimens. The cause can be identified in massive gold scavenging (see Fig. 4.15b and c), mainly caused by the tin contained within the used soldering pastes. This

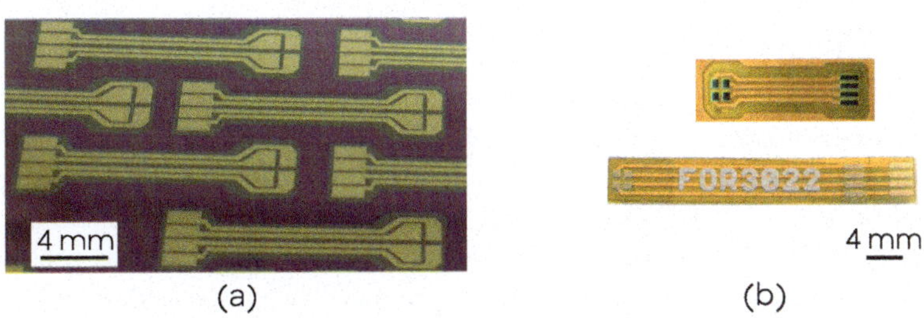

Fig. 4.14 Sensor connection PCB versions. **a** IMSAS cleanroom-manufactured on wafer. **b** Commercially manufactured versions

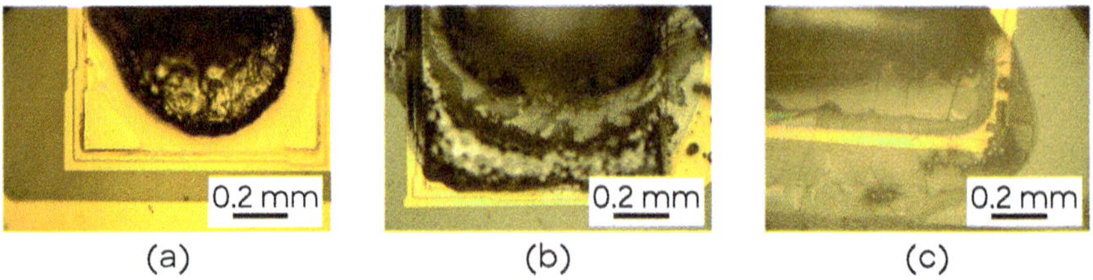

Fig. 4.15 Results of soldering directly on gold pads. **a** Satisfying result using low-temperature soldering paste Almit SnBiAg (139 °C). **b** Gold scavenging using moderate temperature soldering paste Chip Quick SnPb (183 °C). **c** Gold scavenging using moderate temperature soldering paste MG Chemicals SnPb (183 °C)

phenomenon has been reported on in literature already in the 1970s for lead-tin soldering pastes [27] and also more recently in 2018 for one often used type of lead-free soldering paste, SAC305 [28]. The point of failure can clearly be seen in the color differences: adjacent to the solder points, where the gold layer as pad material is supposed to be, the darker substrate material is clearly visible instead. Therefore, for our application this type of ultra thin connection PCB is not directly suited. It could be improved though, to work for the application by, e.g., allowing a thicker gold layer in the pad areas. Investigations on the influence of polyimide inlays on the wave propagation performed by another working group of the research unit, Rittmeier et al., proved a negligible influence of polyimide inlays on the wave propagation [29]. Hence, the requirement to realize the sensor connection PCB to be as thin as possible can be omitted. Therefore, consecutive versions are externally manufactured resulting in a final PCB thickness of 43 μm including metallization using 18 μm thick copper. The final version (see Fig. 4.14b, bottom) also provides an increased distance between sensor and electronics of $d = 36$ mm to prevent reflections to superimpose measurements of the first wave packages passing through the sensor. It also provides an additional set of connection pads for the insertion into a bridge amplifier. With this, it is possible to measure each sensor first with the bridge amplifier, before attaching it to the wireless sensor node.

4.4 Software for Lamb Wave Readout

The sensor node so far is a piece of hardware that can in the following be programmed to gain the function it is supposed to have. For programming purposes, the additional connector on the left PCB side is used, which can be later on cut off to decrease dimensions. The serial wire debug (SWD) programming interface is used which basically only needs two pins for programming and debugging, in addition to supply voltage and ground. A reset connection is also led out to be able to reset the sensor node via the debugger. Two software versions can be used on the wireless sensor node: one simple version which allows a direct sensor data readout at command and one version having a software trigger, that remains in a waiting state until the sensor data contains values above a chosen threshold. The data is then positioned depending on the signal data so that multiple measurements can be compared to each other.

4.4.1 Readout at Command

The software implements basic functions needed to readout either of the two sensor types. If the actuation signal burst is applied in very short time periods, this program is sufficient to sample data from the sensor without any triggering implemented. The advantage of this is a simple, low complexity code with which readout does not depend on any minimum signal level from the sensor. The disadvantage on the other hand is that the signal is not triggered and can therefore not be used for further signal processing like taking the mean measured value over several measurements. Also no direct comparison between two signals is possible without manipulating the signals afterwards. Nevertheless, especially for testing and unknown sensor signal amplitudes or for simple responses from within the material after embedding, to see if the sensor node is still functional, this software version is sufficient. From the start point of the program, which is set directly after the reader field is applied, first the memory addresses are prepared in block sizes that can later on be directly transferred via the communication bus I2C. After that, all active peripherals are initialized by the provided hardware drivers. If a piezoresistive sensor is present as sensor element, a bridge voltage is required, which is provided by a *HIGH* state of the connected GPIO pin. After toggling the pin if required, the I2C bus is initiated, the password is provided for full register access and the energy harvesting feature of the RFID IC is enabled. This has to be done only once per new RFID IC in use as the configuration is retained. Then the ADC sampling routine is started using the DMA peripheral. The DMA causes the ADC data to be directly shifted to memory without being processed by the processor in between. This enables the system to sample at maximum sample rate even if the processor is executing other tasks or running at low speed. For this software version, the memory buffer is filled once, for the software trigger version it is set to circular fill mode, overwriting old values. At this point, for sensor nodes having the software trigger option enabled, it is located next. A detailed description of this function can be found in the next section. If no software trigger is implemented, the measurement data is written to the RFID IC's EEPROM using again the I2C communication bus. After the complete data is transferred, the microcontroller is forced to enter "STOP mode", which significantly reduces the power dissipation and ensures a working HF interface readout of the EEPROM. The readout is prevented during software runtime due to the increased energy requirement, mainly during I2C data transfer.

4.4.2 Software Trigger Solution

An advanced option of software for the wireless sensor node is implemented with a software trigger in between the initializing functions and the transfer of data to EEPROM (see Fig. 4.16). Here, a ring buffer is filled by the ADC and overwritten continuously while the processor compares the values of the buffer to two threshold values. The threshold values can be set via the EEPROM, using memory locations that are not overwritten during sensor data readout. With this, an adjustment of the software trigger is possible after embedding the sensor node into a material. One threshold value is defined above the reference voltage one below. The difference has to be manually chosen to be above the noise level of the expected signal. As soon as the stop condition is reached, the ADC is stopped and the ring buffer is compared value by value, starting from the trigger point toward the start of the buffer. If a sequence of 50 values can be found that are in between two different threshold values, also set via the EEPROM interface, this marks the point in the signal where no Lamb wave information is present. If there are more than three values in a row that are outside this range, it can be assumed that there is some sensor signal present at that point and the counter is reset. Therefore, for very noisy signals it may not be advisable to use the software trigger solution.

After the 50 sequential values close to the reference voltage are identified, the signal is then scanned in the opposite direction for the first positive trigger edge again and a time offset of 50 sampling points is added before the trigger point. This places the start of the Lamb wave signal $t = 1/1.23\,\text{Msps} \cdot 50 = 41\,\mu\text{s}$ away from the left edge when the signal is plotted. This value can be set to whichever value seems fitting for the performed measurement. This is usually done in oscilloscopes as well to shift the signal more into the middle of the screen. With this, the ring buffer is written into another buffer location where additional measurement cycles can be

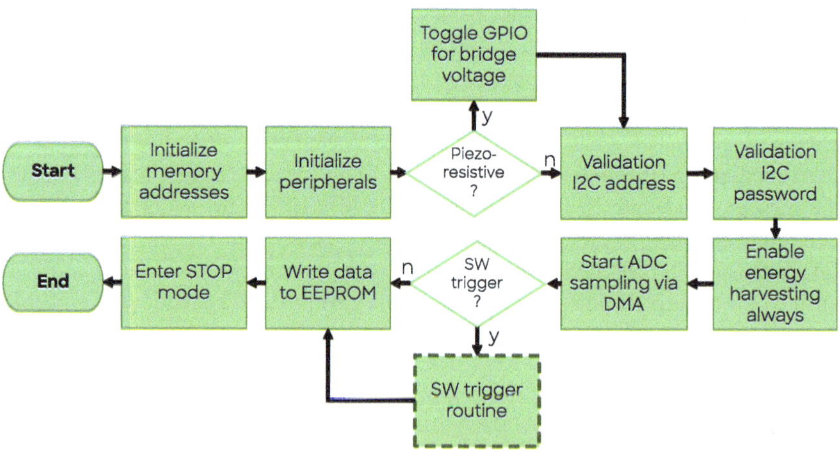

Fig. 4.16 Software flow chart for data sampling and entrance point into trigger routine, if required

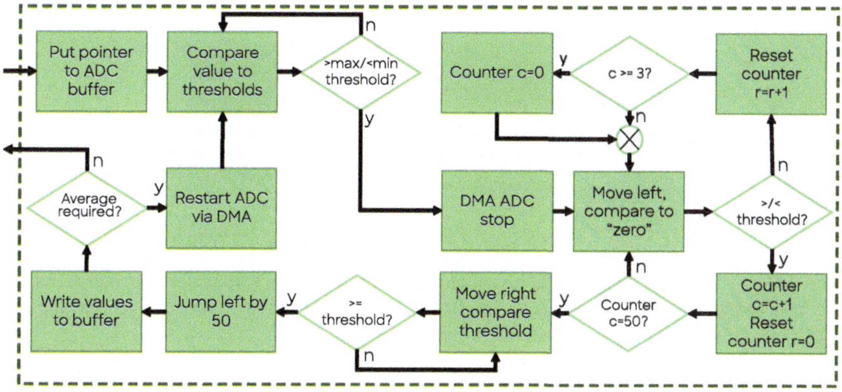

Fig. 4.17 Software flow chart for the software trigger sensor node software

added and an average can be achieved by shifting the result by one bit to the right if required. The flow chart presented in Fig. 4.17 shows the procedure of the software trigger version. The complete C-code containing the software trigger is provided in Sect. A.3, in the appendix of this document. For additional measurement cycles, the ADC is started again and the loop is repeated for the desired repetitions.

4.5 Material Integration of Wireless Sensor Nodes

In the following, considerations for embedding sensor nodes into the target material FML are introduced. They start with possibilities to achieve the thinnest possible height of the complete sensor node. After that the current consumption of the system is investigated for compliance with the chosen wireless approach to supply the sensor node. Lastly, the influence of pressure and temperatures similar to those occurring during the manufacturing of FML on the system is investigated to ensure it can survive the embedding process in the end.

4.5.1 Package Thinning

Most packages of the chosen components have a maximum component height of 600 µm, some are even smaller. The only component exceeding this height is the instrumentation amplifier with a specified maximum height of 800 µm. To reach the minimal possible sensor node height, it will be investigated if the package can be thinned down to some amount while still keeping the IC inside unharmed.

The first step to this is to take a look into the package and inspect, how far the interior of the chip reaches upwards. Especially the bond wires if used for contacting

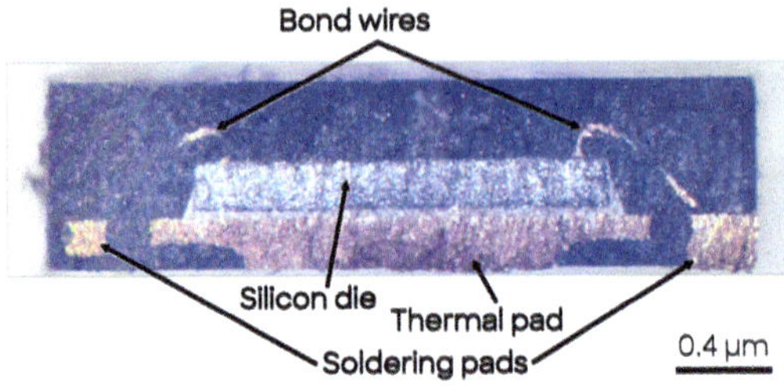

Fig. 4.18 Cut through image of the instrumentation amplifier. The silicon die on a thermal pad can be seen, as well as the soldering pads. The bond wire on the right side is clearly visible in this cut image, the bond wire on the left side is barely visible

inside are expected to reach up higher than anything else. A cut through the package (see Fig. 4.18) shows that the bond wires in fact mark the highest functional point in the assembly at around 150 µm beneath the package surface. Therefore, a careful thinning of the housing by, e.g., mechanical abrasion using a polishing machine, as presented by Giagka et al. [30] or by simply using a hand operated file can be done. It has shown that the removal of about 100 µm results in no damage to the component. End thickness has to be controlled between thinning attempts using a micrometre screw gauge to stop before bond wires are harmed.

4.5.2 *Power Consumption*

To investigate the achieved system performance, a current consumption measurement is done for sensor node version 3 with the simple readout software version. For current consumption measurements, a setup using a laboratory voltage source *Agilent E3620 A* in combination with an amperemeter *Keithley DMM6500* are used (see Fig. 4.19a). The current consumption of RFID IC, piezoresistive sensor and instrumentation amplifier are done using this setup, the value for the voltage regulator is taken from the datasheet. The results of these measurements can be found in Table 4.2.

After this, the complete system with a working microcontroller is measured using the same setup at different clock configurations. It can be seen that for the system running at maximum available internal clock frequency of 16 MHz, measurement and data storage on the EEPROM are done in the shortest amount of time but with the highest power demand of the three configurations of 5.3 mA in average during data transmission via I2C (see Fig. 4.19b). When dialing down the system along with all peripheral clocks to 12 MHz, a very low current consumption of only 2.5 mA in average during data transmission can be noticed for the whole process by only slightly increased run time (see Fig. 4.19c). The drawback here is that as soon as the ADC peripheral is set to 12 MHz, it is only able to work at half sampling frequency. A compromise between both extreme cases can be found in clocking the complete system at 2.097 MHz but leaving the used peripherals at 16 MHz (see Fig. 4.19d).

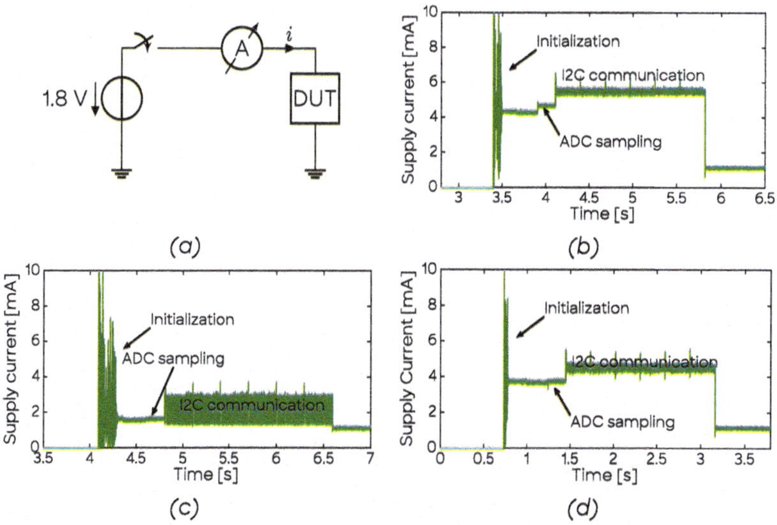

Fig. 4.19 Current consumption measurements for complete system. **a** Measurement setup. **b** All clocks at 16 MHz. **c** All peripherals running at 12 MHz, limited ADC sampling frequency. **d** Complete system running at 2.097 MHz, while ADC set to 16 MHz

Table 4.2 Current consumption of components other than microcontroller

Function	Part-no.	Max. Current
Voltage regulator	TPS7A03	250 nA
RFID IC	ST25DV64K	0.05 mA
Sensor	–	0.25 mA
Amplifier	MCP6N11	1 mA
Total		1.3 mA

The only thing that is going to be complicated by this are operations that have to be done by the system core, like the software trigger, but for the simple software version just measuring, this represents the optimal configuration to reduce the system power demand.

4.5.3 Temperature Stress Investigation

As already introduced in Sect. 2.3, the embedding of the sensor nodes into CFRP-steel is done using elevated temperature, vacuum around the material and seven bar of pressure on top of the material stack to achieve the specified parameters for the FML component in the end. For commercially available prepreg, usually a curing cycle is recommended specifying the manufacturing conditions. The fiber type and

Table 4.3 Curing conditions for a selection of HEXCEL HexPly® prepregs for aerospace application (component thickness <15 mm)

No.	Prepreg type	Max. T [°C]	Max. P [bar]	Duration at max. T [min.]	Matrix material
A	200 [33]	150	7	30	Phenolic
B	913 [34]	125	7	60	Epoxy
C	922-1 [35]	180	5	120	Epoxy
D	M18 [36]	180	7	120	Epoxy
E	M21 [37]	180	7	120	Epoxy
F	8552 [38]	180	7	120	Epoxy
G	F655 [39]	232	5.86	960	Bismaleimide

configuration do not influence this but the resin types within the prepreg come with different curing conditions for different applications (see Table 4.3).

To ensure the developed sensor node will survive the embedding process, three experiments are conducted. They are already published by the author in two publications. The first one investigates the targeted embedding conditions purely on bare components directly out of the box [31]. The second one examines the targeted embedding conditions for four fully assembled and pre-programmed sensor nodes, and the third one evaluates the influence of even higher embedding temperatures on the sensor node's functionality [32].

There are several concerns that provide the reasons for this investigation. The first one is an elevated pressure during manufacturing of FML parts. Even though the selection of components is done with the manufacturing conditions in mind, a maximum of acceptable pressure applied to the ICs, let alone in correlation with elevated temperature, is not specified within any datasheet. The second reason is the possible change of interconnection resistance values for both, active and passive components. This can potentially lead to a different behavior of the circuit after exposure to higher temperatures as reported in [9]. The third concern is a possible degradation of the stored data on the different memory types, implemented within the microcontroller and the RFID-IC. This can be explained looking at the theory of memory degradation under temperature influence introduced in Sect. 4.2. The influence of these challenges during the embedding process will therefore be investigated before an actual embedding is started.

One of the most frequently used resin types is epoxy resin. As Table 4.3 lists, curing conditions for epoxy-based resins can reach from lower temperatures of 125 °C up to 180 °C, while higher temperature is often used especially for load bearing structural parts. Also, one of the used prepregs in this work, the HEXCEL HexPly® 8552 is one of the epoxy-based prepregs cured at 180 °C at 7 bar of pressure. During its complete curing process, the maximum temperature has to be held for 120 min.

First, the individual components are prepared on a metal sheet (see Fig. 4.20). On the left side of the sheet, there is an additional smaller piece of metal with 120 µm thickness, to experimentally validate if the components suffer mechanically

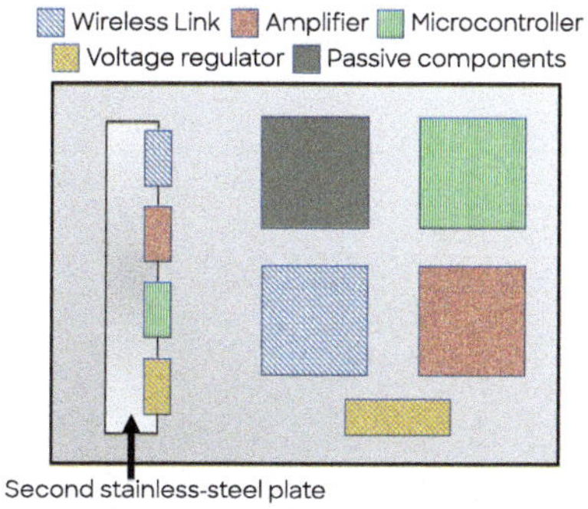

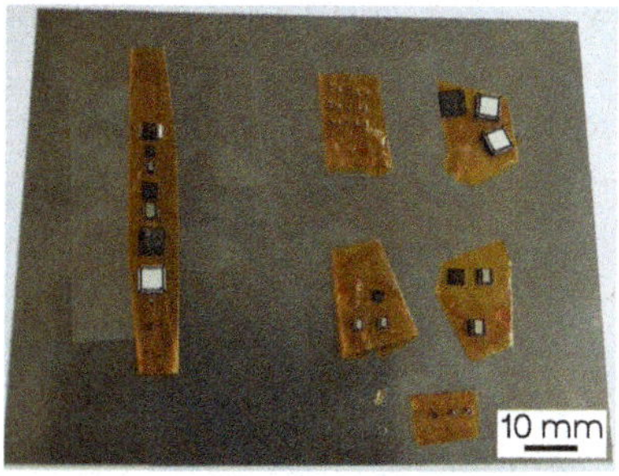

Fig. 4.20 Temperature experiment setup for separate components. **a** Schematic arrangement of components for miniature autoclave experiment. **b** Prepared components, different sides facing upwards

if placed on an edge during elevated temperature and pressure. This is done to simulate localized elevations during the real embedding process due to, e.g., accumulated fibers pressing against the sensor node from the bottom side. The prepared components are then put into a pressure chamber that has also a vacuum connection to act as a miniature autoclave for a simulated embedding process. The complete pressure chamber is put into an oven, slowly heated up, and held at 180 °C for 190 min. After cool-down, the components are extracted, passive components are measured directly and active components are soldered to a PCB for functional tests.

The first result of the pre-tests is that elevated pressure during high temperatures has no effect at all on the components, even if they are directly placed on the edge of a thin metal sheet which is about one fifth of the components' height. None of the components showed any form of deformation. As a result of temperature influence on the passive components it can be established that changes are negligible for resistors and also the changes of the measured capacitors are very small and all in an area according to their original specification (see Table 4.4).

For the active components, the following function tests are conducted:

- Voltage regulator: Boot process, output voltage stability over full input range
- Microcontroller: Boot process, pin-output high/low, ADC, I2C, programmability
- Wireless Link: Boot process, I2C, wireless readout, memory read/write
- Amplifier: Boot process, amplification at $G = 20$.

All required functions are fully executable by every tested component after the experiment. Following this promising result, a full system experiment is done. Four completely assembled and already programmed sensor nodes are put into service and all functions are tested before the start of the experiment. They are then fixated on a stainless steel plate (see Fig. 4.21a), following the same procedure as for the single components before. A temperature sensor, a *Voltcraft TP-202* K-type thermocouple, is placed directly under this plate to measure the temperature directly at the location

Table 4.4 Measurement values for tested passive components before and after the experiment as well as the relative difference

Passive component	Before	After	Difference
Resistor 0402 0.1% 2 kΩ	2.003 kΩ	2.003 kΩ	0%
Resistor 0402 0.1% 2 kΩ	2.003 kΩ	2.003 kΩ	0%
Resistor 0402 0.1% 2 kΩ	2.003 kΩ	2.003 kΩ	0%
Resistor 0402 1.0% 64.9 Ω	65.6 Ω	65.4 Ω	−0.3%
Resistor 0402 1.0% 64.9 Ω	65.3 Ω	65.4 Ω	+0.2%
Resistor 0402 1.0% 64.9 Ω	65.4 Ω	65.7 Ω	+0.5%
Capacitor 0402 X7R 100 nF	93.3 nF	95.9 nF	+2.8%
Capacitor 0402 X7R 100 nF	92.6 nF	95.1 nF	+2.7%
Capacitor 0402 X7R 100 nF	94.7 nF	98.1 nF	+3.6%
Capacitor 0402 X7R 1 µF	892 nF	961 nF	+7.7%
Capacitor 0402 X7R 1 µF	827 nF	943 nF	+14.0%
Capacitor 0402 X7R 1 µF	864 nF	933 nF	+8.0%
Capacitor 0402 X7R 0.47 µF	409.8 nF	417.8 nF	+2.0%
Capacitor 0402 X7R 0.47 µF	417.2 nF	427.7 nF	+2.5%
Capacitor 0402 X7R 0.47 µF	407.4 nF	416.0 nF	+2.1%

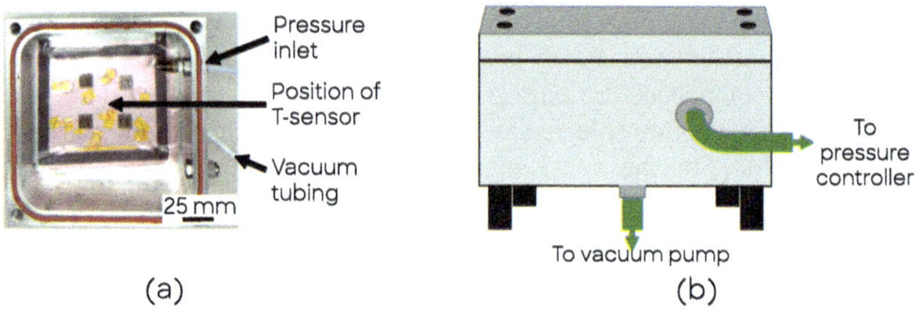

Fig. 4.21 Temperature experiment with complete sensor nodes. **a** Prepared sensor nodes inside miniature autoclave. **b** Miniature autoclave schematic side view

of event. The temperature sensor's wires are lead through the vacuum inlet and exit the vacuum tubing outside the oven to be read out by a *Keithley DMM6500* digital multimeter over time. Again, 7 bar of pressure are additionally applied through the side inlet of the pressure chamber (see Fig. 4.21b) and the complete setup is put into an oven (*Heratherm OMH60-S*) to be heated up to 180 °C temperature. The artificial curing cycle tracked by the temperature sensor inside the miniature-autoclave, depicted by the red, solid line (see Fig. 4.22) covers the recommended curing cycles (A-F, dotted lines) listed in Table 4.3, besides that for bismalemide resin. Temperature measurement shows that the samples are exposed to at least 180 °C for a total duration of 155 min, reaching top temperatures of up to 186 °C temporarily.

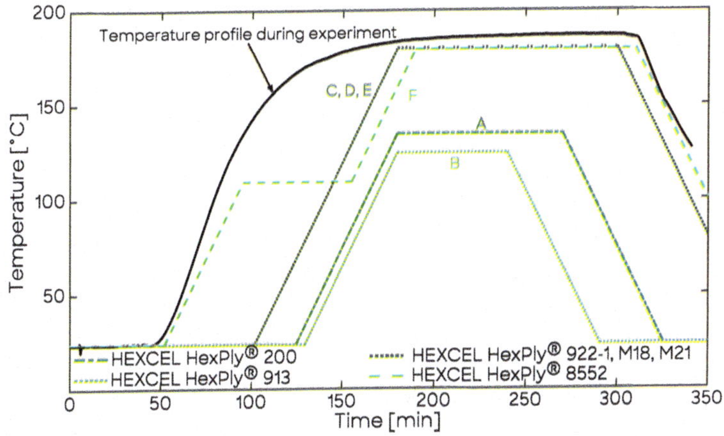

Fig. 4.22 Temperature cycle during the experiment in comparison to recommended curing cycles from Table 4.3

Measurements after the experiment and after a complete cool-down of the sensor nodes show that all four specimens are completely functional. Only a slight color difference of the PCB substrates can be recognized but the material is still intact and does not peel off or similar. Thus, from an experimental point of view, the influence of pressure and temperature on the electronic systems during the embedding process seems to be negligible. Therefore, it can be assumed that they will also survive an actual embedding procedure for epoxy-based FMLs at 180 °C and 7 bar of pressure fully functional.

From a theoretical point of view, there is always an influence happening on the components during elevated temperatures even if the effect cannot be directly recognized by a system failure. The reduction of data retention time specified for all memory components during exposure to higher temperatures generally follows Arrhenius Equation 4.14. With this, a calculation of lifetime reduction during a regular embedding cycle can be done by using an apparent activation energy of $E_{aa} = 1.1$ eV [15]. The embedding temperature cycle starting from 24 °C and ending at 60 °C lasts for 295 min if a medium heat up (2 °C per minute) and cool-down (3.5 °C per minute) are assumed. The curing process is divided into two parts to simplify calculations, the first one presenting the heat-up toward the holding temperature of T = 110 °C for $t_1 \leq 105$ min taken for calculations as constant 110 °C for this time and $105 < t_2 \leq 295$ min at T = 180 °C. This results in an acceleration rate of $A_{F(85,110)} = 10.23$ for the first segment and $A_{F(85,180)} = 1758$ for the second one. This results in a combined lifetime reduction caused by the embedding process of 5591 h, the data retention time still remaining from the specified value of 30 years at up to 85 °C storage temperature [22] is therefore 257,209 h or 29.36 years. With this, it can be noticed that the embedding process at 180 °C reduces the lifetime of the component but it does so only by a small amount of less than one year.

Destructive Temperature Experiment

There are only a few resin types more challenging than epoxy resin on the market. Table 4.3 also lists bismaleimide as one resin type contained in prepregs designed to be used for example for engine parts. This material has a recommended curing cycle of four hours at a maximum temperature of 190 °C and a required post-curing cycle of either 232 °C for 16 h or 243 °C over 6 h. Being an extremely challenging material, it gives the reason for an experimental evaluation of the influence of such conditions, here only regarding temperature without additional pressure, on the sensor node. Taking only the post-curing cycle over 6 at 243 °C hours into account, the lifetime reduction calculated with the help of Eq. 4.14 is found to be 37.5 years, thereby 7.5 years more than the specified value, for the alternative post-curing cycle it is theoretically even reduced by 58.37 years.

By transposing Eq. 4.14 to T_2 the temperature needed to emulate the end of data retention specified, where bit errors might be observable when flash memory is read out can be calculated.

$$T_2 = \frac{1}{-\frac{k_B}{E_{aa}} \cdot ln\left(A_{F(T1,T2)}\right) + \frac{1}{T_1}} \tag{4.21}$$

Using Eq. 4.21, and taking $E_{aa} = 1.1$ eV as before, this temperature can be found to be 501.28 K or 228.13 °C for an experiment time of 10 h. To investigate if this really results in a memory inconsistency, and to simulate similar conditions to the post-curing process for embedding procedures into highly challenging materials like bismaleimide based resins, two sensor nodes are exposed to 250 °C for a total time period of 10 h, including an intermediate measurement after 4.5 h.

At the intermediate measurement, both sensor nodes are allowed to cool down completely first, then an RFID field from the reader is used to power up both nodes for system testing. Results show that the controlled voltage, signal amplification as well as RFID EEPROM readout work as expected, but the measurement of sensor values is not started anymore, thereby the sensor data received by the reader does not change accordingly. Examining the cause of this it can be found that the I2C communication bus is not working anymore, but the microcontroller seems to be still partly working as the GPIO pin is toggled correctly. Both sensor nodes are therefore connected to a debugger and flash memory content is read out and compared to the originally implemented code. No changes can be detected, the content shows to be unaltered. To test if reprogramming solves the malfunction, both sensor nodes are programmed again with the result that by stepping through the software, ADC functionality and data transfer to internal memory can be verified to still function correctly. But for the I2C peripheral not even a clock is generated anymore. It is assumed that although not visible in flash memory readout, the error could be found somewhere within the hardware abstraction layer of the devices but this is not investigated further.

After additional 5.5 h exposed to 250 °C, the tests are repeated and still, voltage regulation is found to be functioning correctly for both sensor nodes, while one sensor node is not showing any harvested voltage from the field anymore. In addition

(a)

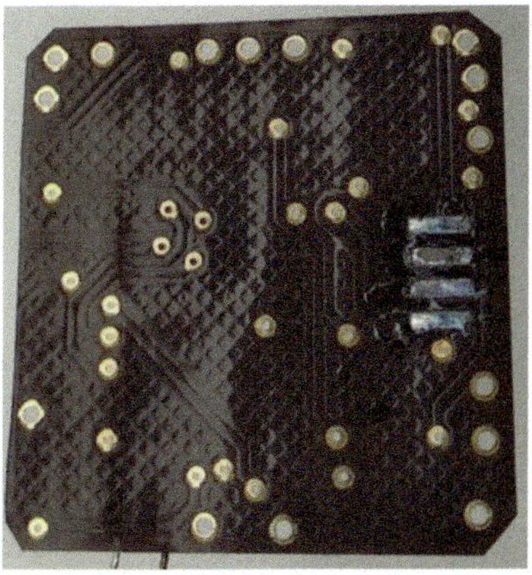
(b)

Fig. 4.23 PCBs after temperature experiment at 250 °C. **a** PCB front view. **b** PCB back view

to that, no readout of the EEPROM is possible therefore, it can be assumed that the temperature caused damage to the RFID IC. The other sensor node is still readable and the energy harvesting functionality is working as expected as well. Apart from the missing functionality at the intermediate measurement, no additional failures are detected. Apart from a decreased functionality, the material used for the flexible PCBs is clearly not designed to be used at such temperatures over a longer duration. Besides a massive change in color (see Fig. 4.23), the flexible top coat is found to be brittle and easily removable from the conductor tracks after the experiment.

This experiment clearly shows that embedding at such high temperatures can not be recommended for the designed sensor node in this work.

References

1. J.P. Lynch, A summary review of wireless sensors and sensor networks for structural health monitoring. Shock Vib. Digest **38**(2), 91–128 (2006)
2. S. Voges, K.-F. Becker, D. Schütze, B. Schröder, P. Fruehauf, M. Heimann, S. Nerreter, R. Blank, S. Gottwald, A. Hofmeister, M. Schmied, M. Kreitmair, W. Fust, R. Neifer, T. Braun, T. Thomas, M. Schneider-Ramelow, K.-D. Lang, Highly miniaturized integrated sensor nodes for industry 4.0, in *IMAPS 2019 - 52nd International Symposium on Microelectronics* (2019)
3. W. Lang, D. Boll, E. Tolstosheeva, A. Ibragimov, K. Schubert, C. Brauner, C. Pille, Embedding without disruption: the basic challenge of sensor integration, in *2012 IEEE Sensors* (IEEE, 2012), pp. 1–4
4. M. Torres Arellano, L. Crouzeix, F. Collombet, B. Douchin, Y.-H. Grunevald, Mechanical characterization of an alternative technique to embed sensors in composite structures: the monitoring patch. Appl. Compos. Mat. **19**(3–4), 379–391 (2012)
5. M. Salas, O. Focke, A.S. Herrmann, W. Lang, Wireless actuation of piezo-elements for the structural health monitoring of carbon-fiber-reinforced-polymers. Mechatronics **34**, 128–136 (2016)

6. F. Berkemeyer, M. Brink, D. Knoop, M. Salas, J.-H. Ohlendorf, K.-D. Thoben, G. Dumstorff, W. Lang, Rotor blade temperature monitoring with material integrated wireless sensor tags, in *Proceedings of the DEWEK 2017, Bremen, Deutschland* (2017)
7. X. Geng, M. Jiang, L. Gao, Q. Wang, Y. Jia, Q. Sui, L. Jia, D. Li, Sensing characteristics of fbg sensor embedded in cfrp laminate. Measurement **98**, 199–204 (2017)
8. K. Dragan, M. Dziendzikowski, A. Kurnyta, A. Leski, J. Bienias, Structural health monitoring of composite structures with use of embedded pzt piezoelectric sensors, in *ECCM-16TH European Conference on Composite Materials* (2014)
9. S. Riches, C. Johnston, Electronics design, assembly and reliability for high temperature applications, in *2015 IEEE International Symposium on Circuits and Systems (ISCAS)* (2015), pp. 1158–1161
10. K.Y. Au, F.X. Che, E.W.L. Ching, Extremely high temperature and high pressure (x-hthp) endurable soi device and sensor packaging for harsh environment applications, in *2017 IEEE 67th Electronic Components and Technology Conference (ECTC)* (IEEE, 2017), pp. 1107–1112
11. P. Singh, D.S. Arya, U. Jain, Mem-flash non-volatile memory device for high-temperature multibit data storage. Appl. Phys. Lett. **115**(4), 043501 (2019)
12. M.C. Morgul, M.N. Sakib, M. Stan, Reliable processing in flash with high temperature, in *2021 IEEE International Integrated Reliability Workshop (IIRW)* (IEEE, 2021), pp. 1–6
13. JEDEC Solid State Technology Association, Jedec standard jesd91b: Method for developing acceleration models for electronic device failure mechanisms (2003)
14. K. Lee, M. Kang, S. Seo, D.H. Li, J. Kim, H. Shin, Analysis of failure mechanisms and extraction of activation energies (e_a) in 21-nm nand flash cells. IEEE Electron Device Lett. **34**(1), 48–50 (2013)
15. I. Inc., An217979: Endurance and data retention characterization of infineon flash memory. Accessed 16 April 2023
16. W. Lang, *Messen in der Elektrotechnik*. Norderstedt: BoD – Books on Demand, 1. auflage ed. (2020)
17. J.N. Haus, W. Lang, T. Roloff, L. Rittmeier, S. Bornemann, M. Sinapius, A. Dietzel, Mems vibrometer for structural health monitoring using guided ultrasonic waves. Sensors **22**(14) (2022)
18. W. Lang, *Sensors and Measurement Systems*. River Publishers Series in Electronic Materials and Devices (River Publishers, Aalborg and Gistrup and Delft, 2019)
19. E. Schrüfer, *Elektrische Messtechnik: Messung elektrischer und nichtelektrischer Größen ; mit 41 Tabellen*. München: Hanser, 9., aktualisierte aufl. ed. (2007
20. Texas Instruments Incorporated, *Analog Engineer's Circuit - Charge Amplifier Circuit* (2022)
21. M. Weber, https://www.mmf.de/instrumentation.htm. Accessed 06 March 2023
22. STMicroelectronics, *ST25DV04K ST25DV16K ST25DV64K, Datasheet, Dynamic NFC/RFID tag IC with 4-, 16-, or 64-Kbit EEPROM, and fast transfer mode capability* (2023)
23. Stm32l031x4/6: Access line ultra-low-power 32-bit mcu arm-based cortex-m0+, up to 32kb flash, 8kb sram, 1kb eeprom, adc, datasheet (2018)
24. Microchip Technology Inc., Mcp6n11 - 500 khz, 800 ua instrumentation amplifier. Accessed 08 Jan 2023
25. STMICROELECTRONICS, Datasheet - tsv7721, tsv7722, tsv7723 - high bandwidth (22 mhz) low offset (200 mv) 5 v op amp
26. M. Salas, O. Focke, G. Stoltenberg, A. Herrmann, W. Lang, Wireless sensor network for structural health monitoring by means of lamb-waves, in *Structural Health Monitoring 2015* (Destech Publications, 2015)
27. F.G. Yost, Soldering to gold films. Gold Bull. **10**(4), 94–100 (1977)
28. D. Hillman, R. Wilcoxon, T. Pearson, P. McKenna, Dissolution rate of electronics packaging surface finish elements in sn3.0ag0.5cu solder. J. Electr. Mat. **48**(8), 5241–5256 (2019)
29. L. Rittmeier, A. Dietzel, J.N. Haus, R. Lammering, A. Mikhaylenko, N. Rauter, T. Roloff, M. Sinapius, Influence of a flat polyimide inlay on the propagation of guided ultrasonic waves in a narrow gfrp-specimen. Special Issue Structural Health Monitoring of Composite Materials, Materials (2022)

30. V. Giagka, N. Saeidi, A. Demosthenous, N. Donaldson, Controlled silicon ic thinning on individual die level for active implant integration using a purely mechanical process, in *2014 IEEE 64th Electronic Components and Technology Conference (ECTC)* (IEEE, 2014), pp. 2213–2219
31. S. Bornemann, W. Lang, Experimental study on stress impact during fml manufacturing on the functional conformity of an embeddable shm-sensor-node, in *The 8th International Electronic Conference on Sensors and Applications* (MDPI, Basel Switzerland, 2021), p. 72
32. S. Bornemann, W. Lang, Considerations and limits of embedding sensor nodes for structural health monitoring into fiber metal laminates. Sensors **22**(12) (2022)
33. Hexcel Corporation, *Hexply 200* (2020). Datasheet
34. Hexcel Corporation, *Hexply 913* (2020). Datasheet
35. Hexcel Corporation, *Hexply 922-1* (2020). Datasheet
36. Hexcel Corporation, *Hexply M18* (2020). Datasheet
37. Hexcel Corporation, *Hexply M21* (2020). Datasheet
38. Hexcel Corporation, *Hexply 8552* (2020). Datasheet
39. Hexcel Corporation, *Hexply F655* (2020). Datasheet

Chapter 5
Structural Health Monitoring

Structural Health Monitoring (SHM) can be seen as a technique inspired by nature. In living organisms, any sort of damage, e.g., a scratch on the skin, is sensed and processed to evaluate the effect and possible actions necessary due to this event. This also applies to an SHM system that is implemented into artificial structures like buildings or vehicles, to achieve the same thing.

In this chapter, a short state-of-the-art for SHM in materials, focusing on ultrasonic methods, is provided followed by the theoretical part as a basis for the presented measurements. After that, the wireless sensor node is used for embedded measurements of Lamb waves in Fiber Metal Laminate (FML). Measurements on and in plates are done for both, piezoelectric and piezoresistive Lamb wave sensors for the intended application. In addition to measurements of undamaged plates to achieve a baseline, pseudo defects in the form of a pair of magnets are used, to validate the functionality of the embedded sensor node system regarding the application in an SHM system. As the focus of this work lies on making sensed Lamb waves readable wirelessly and with this, enabling a complete material integration of a damage detection system, it is proven that both resolution and amplification are suited for this task. Therefore, the assessment of differences due to pseudo defects in the measured waveforms will be done by marking the visible differences in the signal over time. If differences due to pseudo defects can be detected by manual data inspection, the wirelessly transmitted data is expected to be suited for automated damage detection methods as well. For the detection and classification of different damage scenarios, usually, much more advanced techniques like damage detection algorithms based on artificial neural networks are used, which will not be discussed here.

S. Bornemann, *Development of a Self-Sufficient, Wireless Sensor Node Using the Outer FML Layer as an Antenna*, Mechanics and Adaptronics,
https://doi.org/10.1007/978-3-031-70648-6_5

5.1 State-of-the-Art: Material Integrated SHM

This work focuses mainly on embedded SHM systems using Guided Ultrasonic Wave (GUW) and the techniques useful in this context. Multiple detailed explanations of other available methods, like radiographic examination, eddy current testing, and thermographic inspection which cannot be embedded can be found in further literature (e.g., [1–3]) and will not be explained here. Nevertheless, due to the close relation to the method in this work and its application for inspection during maintenance of composite components over the years, a short overview of different ultrasonic techniques using either bulk waves or guided waves will be provided here.

5.1.1 Ultrasonic Waves for Non-Destructive Testing

Ultrasonic testing of materials and especially fiber composite material reaches long back. First developed methods were mainly using bulk waves, while the more recent methods mostly use guided waves.

Water coupled or immersion ultrasonic testing is used as a commercialized, reliable method in aerospace and other industries but due to the elaborate setup, with tanks or jets of water, mainly at the manufacturing stage [4]. To introduce ultrasonic waves into materials and sense them during maintenance, gel coupled transducers can be applied to the surface instead with the drawback that still a coupling agent is required which causes additional costs and a risk of water ingress for moisture sensitive parts [4, 5].

Already in 1998, Castaings et al. suggested to use a technique known as air-coupled ultrasonic scans instead [4]. In recent years, this method gained more and more interest as the resolution increased significantly due to developments in other acoustic applications. This method provides the advantage of working without any additional coupling agent than the surrounding air. Essig et al. introduce different methods of air-coupled ultrasound methods in [6]. If the component is accessible from both sides, the transmission method can be applied. Here, sender and receiver are placed facing each other on different sides of the structure under investigation. While the ultrasound passes as a bulk wave, e.g., through the plate in thickness direction, transmission and reflection losses occur at the boundary layers air/plate-plate/air. Damage in the acoustic path increases the amount of boundaries on the way so a significant amplitude influence can be observed.

Using the air-coupled ultrasonic approach, it is also possible to inspect components when only one side can be accessed. This can be done by two different methods. The pulse-echo method works with the same transducer as actuator and sensor, once again using bulk waves. It is mainly used to inspect surfaces due to a low signal-to-noise ratio for measurements inside plates. Better results can be obtained using the re-emission method, where GUW are excited by one air-coupled ultrasonic transducer and picked up again by another air-coupled ultrasonic probe in a distance [5, 6].

Although air-coupled ultrasonic scans work quite well and are already used for Non-Destructive Testing (NDT) of composite materials commercially, the setup is still not easily applicable in a real airplane inspection scenario, especially during smaller maintenance procedures, where the plane is not completely deconstructed. For this, a material-integrated system, offering the advantage to work without an additional coupling agent by being connected to the material layers at any time would be a significant improvement to any external SHM approaches.

5.1.2 Lamb Waves for SHM Inside Materials

Being an interesting and challenging field of research, the integration of sensors into different materials for integrated condition monitoring has led to a variety of investigations. It reaches from the integration of sensors into civil infrastructure like buildings and bridges to small, bio-compatible sensors inside living organisms. Also, the integration of sensors for multiple purposes into fiber composites has been investigated often in recent years.

For structural monitoring of isotropic materials as well as anisotropic lightweight structures made from Fiber-Reinforced Polymer (FRP) and FML, using GUW in the form of Lamb waves is a very promising approach. They can be introduced into plate-like structures and propagate over large distances with low loss and show a high sensitivity to damage in proximity to their path at the same time. The first description of Lamb waves, which lead to their name, was presented by Horace Lamb in 1916 [7] and has since been used in many works for the detection of damage in plate-like structures. Although this definition was done for isotropic structures, Rittmeier et al. have recently shown in [8] that Lamb waves exist also in FML, and can therefore be used for SHM.

Lamb wave excitation can be done using, e.g., air-coupled ultrasound as presented in [6] or by applying a piezoelectric actuator embedded into or applied on top of the plate. For the latter method, Giurgiutiu et al. have investigated the Lamb wave generation with thin piezoelectric disc shaped Piezoelectric Wafer Active Sensors (PWAS) transducers in both, simulation and experiments in [9]. They found out, that PWAS provide a simple and cost-effective way to generate Lamb waves in plate-like structures which are usable for SHM purposes. Based on this fundamental work, many others have investigated the use of PWAS for SHM purposes as well.

There are several publications in recent years, using embedded piezoelectric transducers for Lamb wave based SHM measurements and comparing them to surface-mounted transducers. Dziendzikowski et al. investigate the differences of surface mounted and embedded piezoceramic sensors in Glass Fiber-Reinforced Polymer (GFRP) finding out that embedded transducers seem to have a better signal-to-noise ratio, which results in an improved damage detection capability when embedded transducers are used for SHM [10]. Feng investigates the measurement capabilities of inlays with inserted piezoceramic transducers within Carbon Fiber-Reinforced Polymer (CFRP) plates with a comparable result, that embedded transducers result in a more reliable localization of the impact damage than surface-mounted ones [11].

Also, real-life applications and data processing of the received sensor signals from networks of PWAS have been investigated frequently in recent years. On the application side, Senyurek successfully investigates the use of Lamb waves for cut- and impact damage detection on Boeing 737 wing parts in [12]. As an example for data processing, Hameed et al. [13] use an active, wire-bound Lead Zirconate Titanate (PZT) sensor network to detect, localize and quantify the size of damage in an aluminum plate in a multi-stage approach that does not require a baseline measurement. The signals are sampled over 200 µs at a sampling rate of 1 GHz and averaged over 500 measurements for noise reduction.

Many publications can be found which describe the influence of real and artificial defects on the wave propagation behavior of Lamb waves in isotropic and anisotropic materials [14–16]. An additional way to introduce a changed Lamb wave response to a plate, being a fully reversible approach is presented by Pfeiffer et al. in [17], who are using pseudo defects closely pressed to the plate, observing similar changes in the sensor response as for certain real defects. This technique is used in form of a pair of magnets, placed on both sides of the plate, in this work as well.

Despite increased research in the area in recent years, no material integrated system that is able to work wirelessly, amplify, digitize, and roughly process the sensor data received for Lamb wave based SHM in FML, besides the one presented in this work and the associated publications, is reported in literature so far.

5.2 Theoretical Background: Wave Propagation

As already introduced in the previous section, ultrasonic waves traveling as guided waves through plates or plate-like structures can be used to detect possible flaws and damage in, e.g., airplane parts. The correlation of wavelength, velocity, and frequency inside materials follows the same relations as that for air as already used in calculations for electromagnetic waves, using Eq. 3.5. The phase velocity has to be adjusted to that of the material. For excitation frequencies resulting in a wavelength greater or in the order of the plate thickness, GUWs are introduced, which can travel large areas while only losing little energy on their way along the plate. Given Eq. 3.5, this means thinner structures can generally be investigated at higher frequencies using GUW based techniques than thicker structures of the same material.

In the case of plate-like structures, i.e., structures with two free surfaces, being the outer areas of a plate surrounded by air, two generally different types of GUW can be excited [18]:

- Shear horizontal waves
- Lamb waves

The first type of waves, as already indicated by their name, propagates in parallel to the plate's surface while an in-plane particle motion perpendicular to the wave propagation direction can be observed [18]. Shear horizontal waves do not result in significant out-of-plane displacements, especially useful to detect, e.g., a delamination between layers in case of FML, but can be used for other SHM applications nevertheless.

The main focus of this work lies however on the use of the second type of GUW: Lamb waves. In contrast to shear horizontal waves, Lamb waves show a significant displacement in thickness direction of a plate or plate-like structure and can be used for SHM applications as well. In-depth, theoretical descriptions of Lamb waves and their characteristics can be found in numerous books (e.g. [18–20]) many of them having an SHM background. They will therefore be described here only briefly in the most important aspects for this application.

Lamb waves come in two main modes when moving through a plate. They can be separated into symmetric, "SX" and antisymmetric "AX" modes, where the X represents the wave mode order. The fundamental modes S0 and A0 are usually preferred for SHM purposes. Higher order modes occur at higher frequency and plate-thickness combinations which can complicate the analysis of the sensed Lamb wave signal. Therefore, thin plates in combination with lower actuation frequencies are preferred for most damage detection applications. Figure 5.1a shows the difference between symmetric (top) and antisymmetric (bottom) waves in a plate element. It can be seen that both waves can be measured on top of the plate, while for an integrated sensor close to the middle of the plate, the symmetric mode influence becomes much smaller than the antisymmetric one regarding displacement amplitude in z-direction.

In addition to this, Lamb waves are dispersive waves which means their phase and group velocities depend on the frequency and on the characteristics of the medium through which they are propagating. Dispersion diagrams show the correlation of plate thickness, phase velocity, group velocity, and wave number for a given plate material and are required for an understanding and interpretation of experimentally

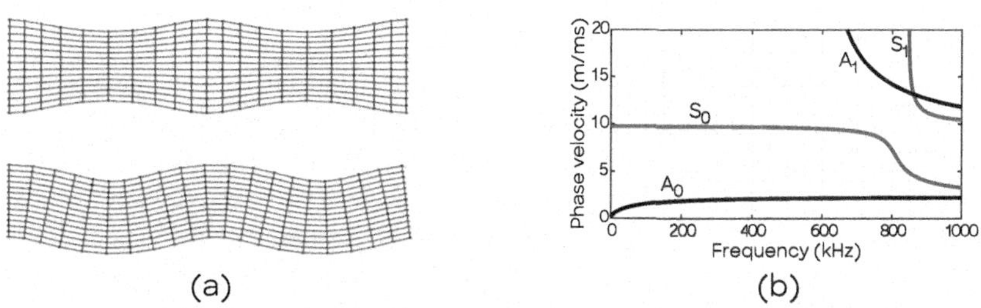

(a) (b)

Fig. 5.1 Lamb waves in plate-like structures. **a** Z-component of motion for Lamb wave modes that can be used for damage detection [21]. Top: Symmetric mode. Bottom: Antisymmetric mode. **b** Exemplary dispersion diagram for a pure 2 mm thick unidirectional CFRP plate, from the predefined CarbonEpoxy2_Rokhlin_2011 material configuration, showing higher order wave modes at increased frequencies, generated using Dispersion Calculator V2.1

obtained measurement data. These characteristics for a given material can be calculated using the following relations (presented here for isotropic, elastic plates with stress-free boundaries). The characteristic equations for Lamb wave motion in plates can be expressed by the Rayleigh-Lamb equations [22–24]

$$4k^2 pqsinf(ph)cos(qh) + (k^2 - q^2)^2 sin(qh)cos(ph) = 0 \tag{5.1}$$

for symmetric waves and by

$$4k^2 pqsinf(qh)cos(ph) + (k^2 - q^2)^2 sin(ph)cos(qh) = 0 \tag{5.2}$$

for antisymmetric waves. With

$$p^2 = \frac{\omega^2}{c_L^2} - k^2, q^2 = \frac{\omega^2}{c_T^2} - k^2 \tag{5.3}$$

with $k = \omega/c_p$ as the wave number, the angular frequency $\omega = 2\pi f$, the phase velocity c_p, the longitudinal velocities of the material c_L and c_T as the transverse velocities of the material, respectively. Transformed to c_p they result in frequency-dependent phase velocities, which can be plotted over frequency and thereby result in dispersion diagrams as the one depicted in Fig. 5.1b. The group velocity with which the wavefront of the modes propagate through the material can be calculated by [22, 24]

$$c_g = \frac{c_p^2}{c_p - f\frac{\partial c_p}{\partial f}}. \tag{5.4}$$

For a quick calculation and graphical display of dispersion diagrams for a known material, the open-source, *MATLAB*-based *Dispersion Calculator*, developed and provided by the *German Aerospace Center (DLR)* can be used. The exemplary phase velocity dispersion diagram shown in Fig. 5.1b is calculated for a 2 mm thick pure CFRP plate with material data for "*CarbonEpoxy2_Rokhlin_2011*", a material already implemented in the database.

The dispersion diagram shows that for this material and the chosen thickness only the fundamental Lamb wave modes S0 and A0 are found for excitation frequencies lower than 650 kHz while the S0 mode propagates much faster through the plate than the A0 mode. At higher frequencies, the phase velocity of the S0 mode reduces to the same range as that of the A0 mode but higher order modes S1 and A1 are excited with higher phase velocities at the same time. This makes a superposition of the different wave packages much more likely and by this usually damage evaluation more complicated.

5.3 Material Integrated GUW Measurement Unit

For the measurements presented in this chapter, both system components introduced in the previous chapters are combined and present the wireless sensing system for FML material integration. The complete system can be seen in Fig. 5.2. It consists of an antenna that is attached to the plate after manufacturing, an embedded electronic sensor node made from commercial components inside the plate (visible due to a cutout milled into the final plate), and an attached sensor on the sensor node which can be either piezoelectric or piezoresistive.

5.3.1 Material Integration Procedure

For the first material integration tests, version 2 of the sensor node is embedded into a GLAss REinforced laminate (GLARE) plate with two aluminum layers, one at the top and one at the bottom of the plate. The embedding is done at 120 °C over two hours.

The idea behind this experiment is to investigate if a layer of Ultra Violet (UV) curing glue of type *DELO® KATIOBOND® GE680* around the components is on one hand mechanically stable enough to keep the components in place during the embedding process and provides sufficient electrical insulation against the metal layers of the stack-up on the other hand. For this, four sensor nodes are prepared with a software Pulse-Width Modulation (PWM) signal generated at their output depending on the provided input at the sensor interface, so that the main component functions can be verified. The sensor interface has the pre-versions of the later on used piezoresistive sensors attached to them. As they are not encapsulated and at this point work as pure pressure sensors, no variation of the signal after embedding is expected, but an output signal should be visible nevertheless. The output and power supply of the nodes is lead to the outside using wires for this experiment. Two of

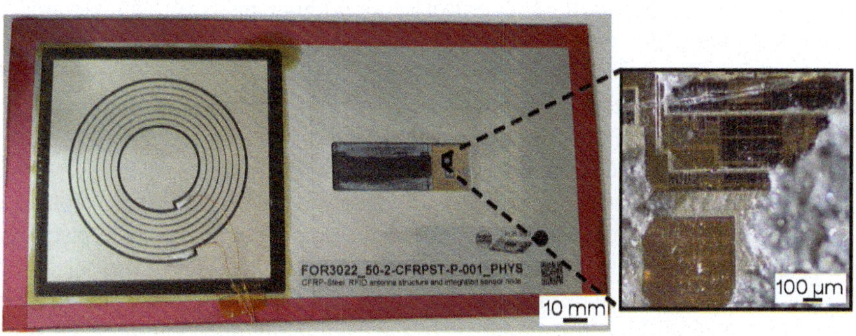

Fig. 5.2 Antenna and embedded electrics in plate, milled cut to show the embedded sensor node inside the plate. Right side: Microscopic zoom of one cut-open IC

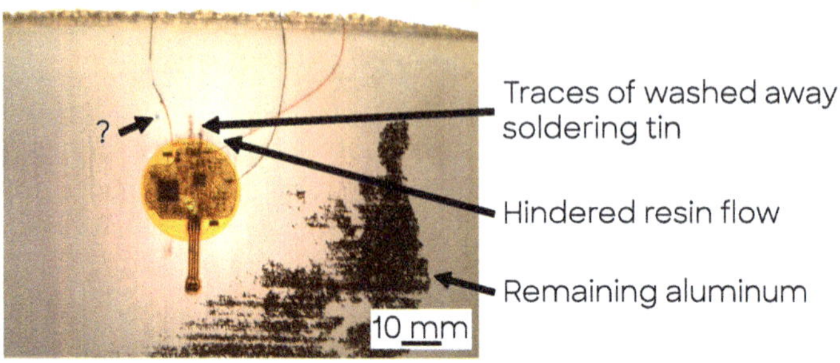

Fig. 5.3 Transmitted light image of the embedded sensor node inside GFRP without any protection by UV curing glue

them are covered carefully with UV curing glue on all conductive parts of the Printed Circuit Board (PCB) while two are left without any additional insulation.

The results show that one of the UV glue covered sensor nodes survived the embedding process and shows the expected output signal after supplying power to it. The other UV glue covered one shows no output signal, as well as both of the uncovered nodes. To examine what happened to the sensor node components during the embedding process and if components got somehow removed from the PCB, the outer aluminum layers are removed by wet chemical etching, uncovering the GFRP layers with one of the unprotected sensor nodes inside (see Fig. 5.3).

It can be seen, that the sensor node components seem to have remained in place. Above the integrated element, there are traces in the form of the sensor node that could give a hint that the resin was not completely able to surround the sensor node at this point. It could also be explained by a slightly moved node during the embedding process, being dragged toward the bottom of the image. This would also explain the dark traces that can be noticed in this area above the sensor node. In this region, it is assumed, that some of the soldering tin has been washed away during the embedding process, although process temperatures stayed well below the melting point of the used soldering paste. Nevertheless, this could explain the malfunction after embedding observed for both sensor nodes without any covering layer around them. Another remarkable part is the area marked with an arrow and a question mark. There seems to be some kind of square-shaped foreign body inside the GFRP which should not be present there. From the visual inspection possible, given the image, all components on the PCB are still present, therefore it is assumed that this is just a foreign body introduced to the laminate during manufacturing.

Nevertheless, for further embedding experiments, it can be noted, that a protective layer of some kind is required. For one thing, to keep all components in place and keep the soldering paste where it is supposed to be. In addition to that, electrical insulation surrounding the sensor node is required anyway. In GLARE, to avoid short circuits toward the aluminum layers and even more important in CFRP-steel, because of the conductive fibers. Therefore, all sensor nodes to be embedded are covered with a thin layer of epoxy film adhesive of type *LOCTITE EA 9396 AERO* in the upcoming material integrations. This is done using a vacuum setup inside an oven at 120 °C for curing while keeping the film adhesive in place with a mold made of PolyTetraFluorEthylen (PTFE) sheets. This also prevents the attached sensor to be covered with epoxy during the pre-embedding.

5.3.2 Data Readout

For wireless sensor node supply and data readout, an Radio-Frequency IDentification (RFID) reader unit is used. The *ST25R3916-DISCO* is a development board and offers a graphical user interface with which it is possible to read the large Electrically Erasable Programmable Read-Only Memory (EEPROM) of the RFID Integrated Circuit (IC). The values can be exported as a binary (.bin) file and stored. Using the following commands in *MATLAB* stores the former binary values into a column vector "var" of type double which can then be used for data processing or to generate a plot of the raw data. The input data format has to be chosen accordingly. As an 8 bit resolution is chosen for the Analog-to-Digital Converter (ADC) and are kept in that format when stored into the EEPROM, "uint8" is used.

```
fid = fopen(['path\name.bin'],'r');
var = fread(fid,Inf,'uint8');
```

The analog voltage values that are represented by the stored data can be calculated by

$$v_i = \frac{var_i}{255} \cdot 1.8\,\text{V}. \tag{5.5}$$

whereas the time difference between two data points can be gained from the known sampling rate f_s, determined by Eq. 4.16 of 1.23 Msps using

$$\Delta t = \frac{1}{f_s} \tag{5.6}$$

5.4 Measurements with the Wireless Sensor Node

In this section, wireless measurements performed with the wireless sensor node are presented. It starts with introducing the actuation of Lamb waves during the measurements and continues with measurements inside plates with the piezoelectric sensor and magnets placed as pseudo defects. After that, measurements performed using the piezoresistive sensor are presented as well. For the piezoresistive sensor, measurements with applied as well as integrated sensors are presented.

5.4.1 Actuation of Lamb Waves

For the measurements of GUW by wireless integrated sensor nodes, actuators of type *PRYY+0237* from manufacturer *PI-ceramic*, with a diameter of 16 mm, a thickness of 0.5 mm and of material type *PIC255* are applied on top of the plates. Comparisons of different types of coupling agents for a surface-applied actuator show a significant influence on the signal amplitude measured inside the plate with permanent gluing being the best option in the comparison [25]. Therefore, this method is used in the presented results as well. For actuation, a hann-windowed sine burst (Fig. 5.4) is applied to the actuator using an *Agilent 33522A* Arbitrary Waveform Generator (AWG) whose signal is then amplified by 50 using a *WMA-300* high voltage amplifier. The advantage of using a hann-windowed sine burst is that significantly reduces the excitation of frequency side lobes around the targeted center excitation frequency [9].

5.4.2 GLARE Plate, Piezoelectric

For Lamb wave measurements with the developed system, a GLARE plate is manufactured with two embedded wireless sensor nodes and one piezoelectric sensor attached to each electronic. The square plate has an edge length of 300 mm and is made of 16 GFRP layers between two aluminum layers and is thereby 2.88 mm thick. The software trigger functionality is implemented on the microcontroller. This will allow for measurements that are aligned at the same point and are thereby much easier to be compared to each other.

One sensor shows full functionality after the embedding which is sufficient for the plate to be used for GUW measurements. The schematic plate setup is shown in Fig. 5.5a, while a photo of the plate with marked sensor node position and surface-applied actuators can be seen in Fig. 5.5b. For different measurement scenarios, actuators are applied at four different locations on the plate's surface. For the results presented here, the actuator aligned directly facing the sensor in fiber direction, the one with green and yellow wire attached, is used. The actuators are permanently

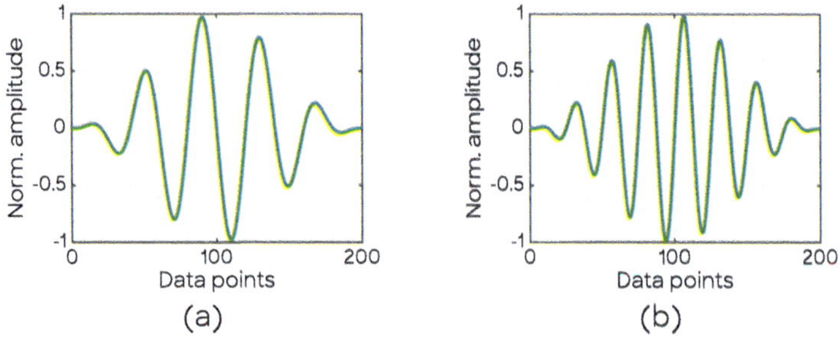

Fig. 5.4 Hann-windowed sine burst signal fed into the AWG. **a** 5-pulse burst. **b** 8-pulse burst

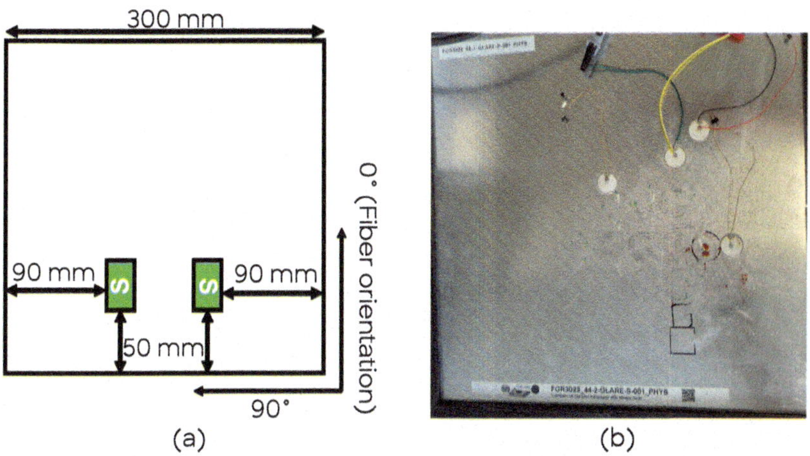

Fig. 5.5 GLARE plate with piezoelectric sensors on sensor nodes. **a** Schematic representation of the placement of embedded nodes. **b** Photo of the actual plate with marked sensor node position and surface-applied actuators

glued to the plate's surface to ensure a comparable Lamb wave actuation even if measurements are performed again at a later time. As the edges of the plate have the potential to cause Lamb waves to be reflected back to the center of the plate, this effect can be reduced by applying tape, in this case, vacuum tape around all edges. To keep the tape from sticking elsewhere and improve handling of the plate, it is put into a surrounding 3D-printed frame, lifting it above the table and improving handling when it comes to the application of pseudo defects.

The antenna connection is led out of the plate's edge using enameled copper wire and connected to an antenna outside the plate itself to be able to evaluate both systems separately before connecting them.

Measurement Results, 50 kHz with Pseudo Defects

In the finished GLARE plate, measurements with magnets as pseudo defects are performed. The magnets are round ones with a diameter of 20 mm and a thickness of 5 mm and are placed between sensor and actuator according to Fig. 5.6. They are always placed as a pair, one on top of the plate and one below, placing just one magnet on the plate does not show any effect. The measurements are performed with a peak-to-peak actuator supply of voltage of 30 V at a Lamb wave center excitation frequency a 5-cyclic Hann-windowed sine burst of 50 kHz as with this frequency differences in the wave arriving at the sensor are already well visible. Later on, also measurements at higher frequencies are shown in comparison.

Considering reproducibility of the measurements it is important that when the pseudo defects are moved to another place on the plate, they are not just pushed to

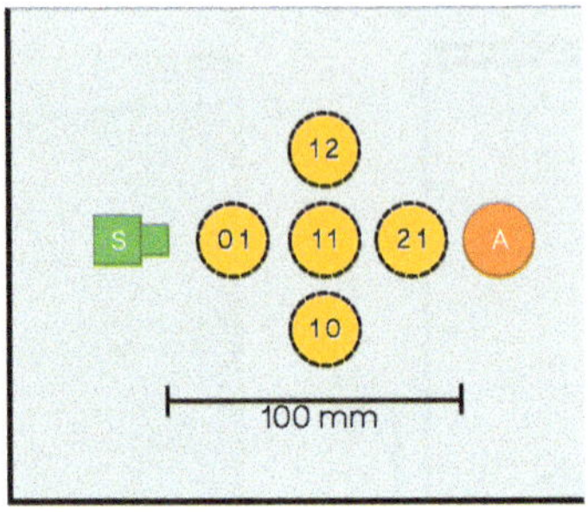

Fig. 5.6 Zoom-in on one corner of the plate, positions of the magnets (yellow, coordinates XY) placed on the plate as pseudo damage, position of actuator (orange, A) and position of the embedded sensor node with sensor (green, S)

the side but slightly lifted before moving. This is to prevent scratches in the material and thereby small particles forming bumps between the magnet and plate surface which can cause local spots of increased pressure to the plate and thereby change the result from one measurement to the next. As long as this is taken care of, repeated measurements of pseudo defects at defined positions as well as repeated baseline measurements show very good reproducibility.

For the first investigation, several baseline measurements are repeated each lying two hours apart on the same day. The results show very good compliance with the course of the previous curves (see Fig. 5.7). Beginning from 250 µs there are slight changes occurring in the amplitude of the curves but the general shape stays unchanged. As already mentioned, the measurements are performed on the same day at the same location and environmental conditions. It can be expected that under changed conditions, e.g., a different temperature, the measurements look different. As this is not the focus of this work, all measurements are performed roughly at room temperature, and bigger temperature changes are not considered. To train later on damage detection algorithms this has of cause to be taken into account.

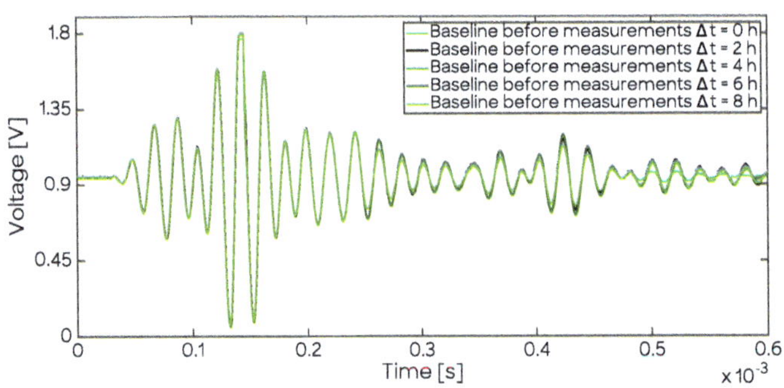

Fig. 5.7 Baseline measurements distributed over one day at room temperature before any magnets are placed on the plate

In the next step, magnets are placed on different positions between actuator and sensor as shown in Fig. 5.6 by the yellow circles containing coordinates of placement for referencing. The different positions and the changes of the measured waveform will be presented here one by one in the following.

Starting at the leftmost position "01", closest to the sensor, the pseudo damage measurement compared with the baseline measurement shows clear differences. The points where the most significant changes are visible are additionally marked with arrows. It can be seen that the first wave package is only slightly influenced while already in the second visible wave package a significant damping of the Lamb wave occurs. The same appears for the region after the second wave package by which a much more separated wave package around 250 µs emerges. The last clearly visible wave package from the baseline measurement is almost completely damped by the application of magnets. It can be confirmed that the method of applying magnets to both sides of the plate results in changes to the wave propagation behavior in a plate-like structure which will be used here to validate the capabilities of our measurement system to differentiate between undamaged plate, plate with pseudo damage applied and pseudo damage at changed locations.

If the magnets are moved to position "11", in the middle of the signal path between sensor and actuator, the measurement result looks different in comparison to the baseline (see Fig. 5.9) and also to the measurement shown in Fig. 5.8 at position "01". It can be noticed that the first and second wave package get even more damped than before. At 350 µs a phase shift occurs, probably caused by two superimposed wave packages. The last section after 400 µs shows significantly less damping compared to the measurement with magnets at position "01".

Moving the pseudo defect even closer toward the actuator, to position "21", results in a less damped first wave package, while the second one is even smaller in amplitude than for the previous measurements (see Fig. 5.10). A small phase shift is visible at 200 µs and between 300 µs and 400 µs the signal amplitude becomes very small while also a phase shift of about 90° is visible at 330 µs. It is assumed that this is again caused by superimposed waves. After 400 µs the signal is again in phase with the baseline signal, and almost as much damped as in the measurement shown in Fig. 5.8.

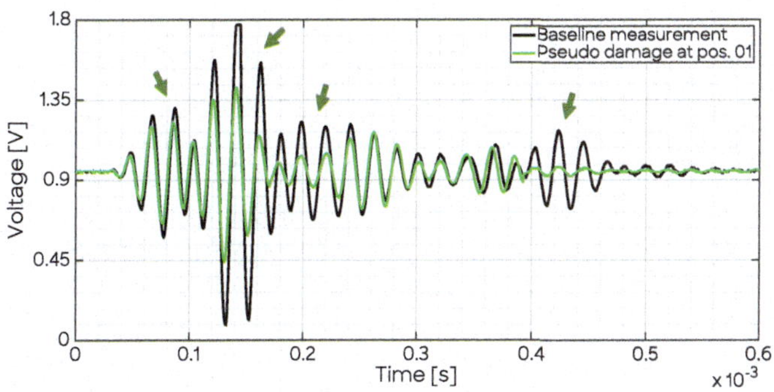

Fig. 5.8 Pseudo damage at position "01" compared to baseline

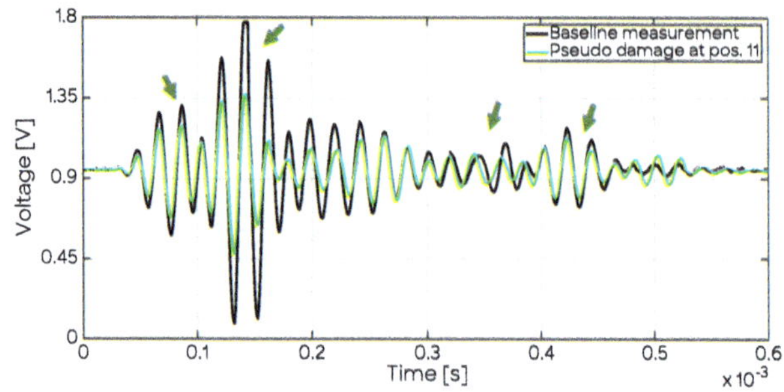

Fig. 5.9 Pseudo damage at position "11" compared to baseline

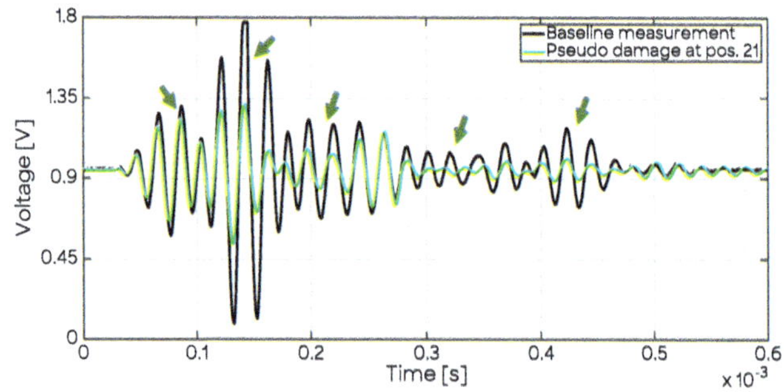

Fig. 5.10 Pseudo damage at position "21" compared to baseline

When moving out of the measurement path, to position "10", the characteristic signal influence reduces. The first wave package is completely in accordance with the baseline measurement while the second one shows a slight increase at its 4th and 5th cycle. In the following cycles where no clear wave package can be defined, also a slight reduction in the signal can be observed, while there are no significant phase shifts visible anymore. For the last visible wave package, a very slight reduction of the signal compared to the baseline can be observed, but the signals do almost share the same amplitude again (Fig. 5.11).

At the other side of the pathway between sensor and actuator, the signal looks different as well but also here the relative changes are much smaller than with magnets applied directly on the path. Actually, the first and second wave packages are completely congruent while after the second one, a slight reduction is visible here as well. Also starting from around 300 µs, phase shifts due to superposition of waves are again visible together with an attenuation of the amplitude (Fig. 5.12).

For a pseudo defect position even a bit further away from the measurement path, almost no signal influence can be detected anymore. Position "13" can be found directly next to position "12" moving away from the sensor-actuator path. The signal with magnets as pseudo defects only differs at two points slightly from the baseline signal (see Fig. 5.13). At the first point, a very small phase shift as well as an attenuation can be observed while at the second point, an increase in the small wave

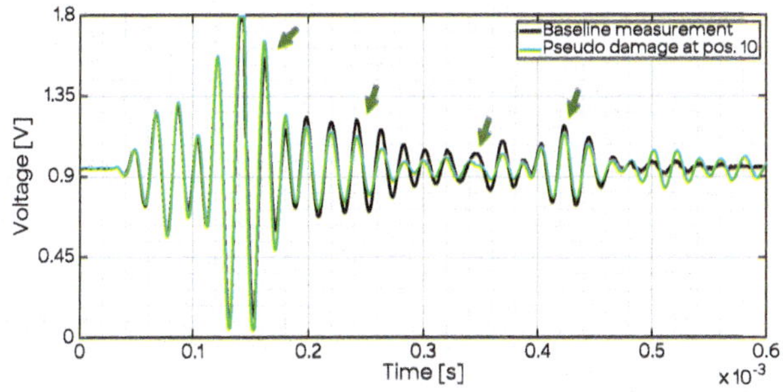

Fig. 5.11 Pseudo damage at position "10" compared to baseline measurement

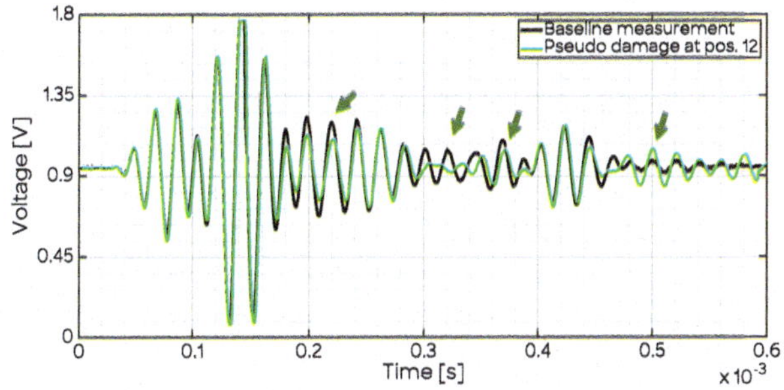

Fig. 5.12 Pseudo damage at position "12" compared to baseline measurement

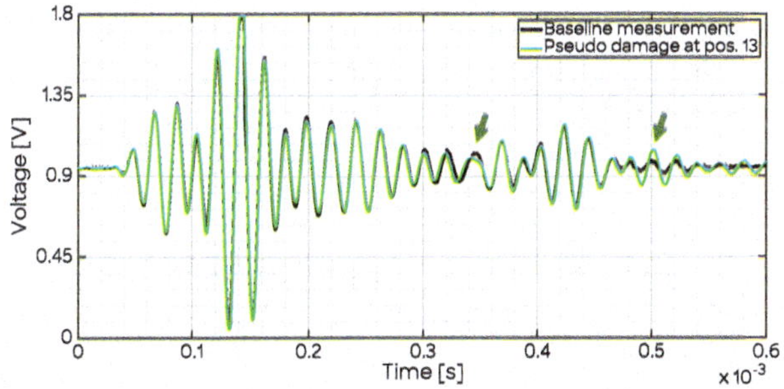

Fig. 5.13 Pseudo damage at position "13" compared to baseline measurement

amplitudes can be seen. The second point may already be some kind of edge reflection or similar, less important for damage evaluation.

Therefore, it can be said that if a pair of magnets as pseudo damage with the presented geometry is applied close to the measurement path, it can be detected in changes in the received signal with the introduced material integrated system. The system provides a sampling rate and resolution that is sufficient to detect this type of pseudo damage quite reliable. The next question is if the baseline signal is the same after completed pseudo damage measurements as before the measurements.

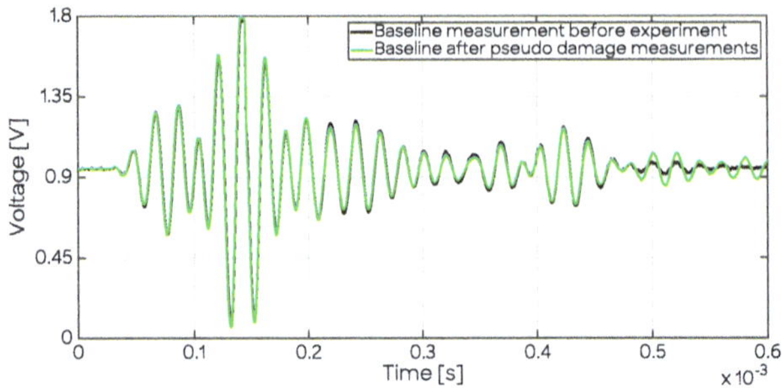

Fig. 5.14 Baseline measurements before and after the experiment with pseudo damage placement at room temperature

Therefore, as a concluding measurement to this experiment, the baseline measurement obtained after all measurements with attached magnets are done is once again compared to the baseline obtained before the start of measurements. It can be seen in Fig. 5.14, that the measured curves are once again in good alignment. Both signals are in compliance and no significant amplitude differences or phase shifts are apparent. The changes in wave propagation caused by applied magnets appear to be completely reversible once the magnets are removed again.

In addition, this also means that the presented system is capable of resolving small differences in the signal which can be a hint of structural damage. With this, the system is fully suited to digitize and transfer piezoelectric sensor data for material-integrated Lamb wave measurements for SHM purposes. The next consecutive step toward fulfilling the given requirements from the beginning of this work is to prove the exact same thing for a piezoresistive sensor approach as well.

5.4.3 Piezoresistive Sensor: Preliminary Experiments

Measurement on a Glass Slide

Before embedding a sensor node of version 3 with a piezoresistive sensor attached to it into a plate, a first validating experiment using a setup on a glass slide in a distance of 30 mm from the actuator is performed (see Fig. 5.15a). The actuation is done using a 75 V peak-to-peak voltage, the center excitation frequency is varied between 30 kHz and 200 kHz. The signal is recorded with an amplification set to $G = 473$, based on the maximum achieved signal to system voltage. This is an important relation because the system voltage determines automatically the maximum ADC voltage and thereby the resolution of the system. This maximum can be found at the first resonance frequency of the given sensor of 50 Hz [26]. If the center excitation frequency is chosen below or above this value, the maximum signal amplitude after digitization is significantly smaller than the complete operating range (see Fig. 5.15b).

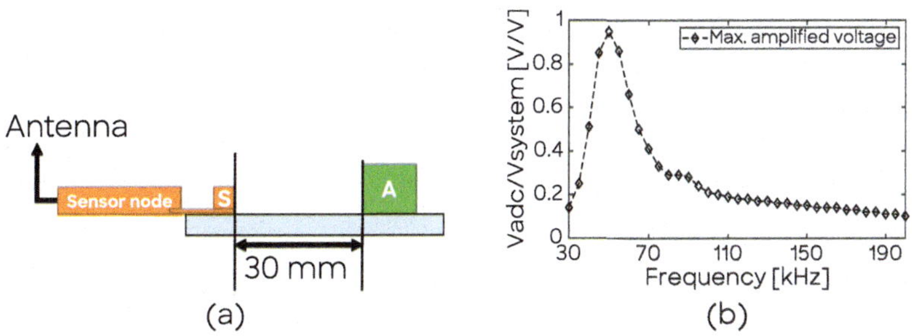

Fig. 5.15 Measurement of the piezoelectric sensor on a glass slide. **a** Schematic measurement setup. **b** Resulting in sensor signal to system voltage amplitude over frequency for a 75 V_{pp} actuation with a gain of G = 473

This means that the resolution of 8-bit will not be used in most cases which means the system will not work with its best possible performance. Therefore, to work with this type of sensor, the working bandwidth has to be defined beforehand to be able to adjust the amplification to a fitting value for the expected signal amplitudes. Otherwise, to cover a large bandwidth including the resonance frequencies would require a system providing a higher resolution to still resolve small amplitudes occurring between the resonance frequencies of the sensor. As the second option can not be implemented that easily by several system limitations, the restriction of the measurement bandwidth will be the method of choice. Looking at the provided transfer curve this means that the given amplification is still not high enough to operate the sensor node with attached piezoresistive sensor outside the resonance frequency.

At this point, it can be noticed that the slowly decreasing voltage ratio is not caused by the sensor or at least not by the sensor alone. Given Eq. 4.11 and an Gain-Bandwidth Product (GBWP) of typically 35 MHz, the set amplification of G = 473 can only be provided by the amplifier up to a frequency of $f \approx 74$ kHz. For higher frequencies, a lower amplification than the set one will be achieved. This point can be clearly seen in the transfer graph (see Fig. 5.15b). It can be found slightly higher for the given amplifier, at around 85 kHz. For further increasing frequencies, a significant decay can be seen. Nevertheless, it is assumed that with proper amplification, this sensor type attached to the wireless sensor node will also be suitable for Lamb wave detection and will then provide a lead-free, robust, and functionally compliant alternative to PZT sensors. This is shown by measurements performed with the improved, double-amplifier sensor node version 4, presented in Sect. 5.4.4.

System Integration Using the Autoclave Process

After promising results from the preliminary embedding experiments presented in Chap. 4 and glass slide experiments, the system integration into a plate made from CFRP and stainless steel, produced at 180 °C inside the autoclave at the *Institute of*

Mechanics and Adaptronics (ima) in Braunschweig is investigated. Again, a piezoresistive sensor is attached to the sensor node with a single amplifier version 3 and an amplification of G = 473. This represents the embedding experiment under real manufacturing conditions in an autoclave after extensive preliminary tests outside the laminate. As anticipated, the sensor node shows complete functionality after the embedding process.

To test the functionality of the sensor node after embedding and compare the results to those obtained on the glass slides, an actuator is applied on top of the plate, 70 mm away from the sensor location. The measurements show that the embedded system is still functional after the embedding and that for 50 kHz, the signal amplitude is sufficient for readout. For higher frequencies, above the first resonance frequency of the sensor, the amplitude becomes very small and cannot be resolved anymore. Therefore, for measurements with piezoresistive sensors an improved version of the sensor node is developed which provides a double-amplifier stage, offering sufficient total gain for the application. Results of the measurements performed with this version are presented in Sect. 5.4.4 (Fig. 5.16).

Regarding antenna integration, a quadratic area is cut out of the upper metal layer before manufacturing to allow for the placement of the antenna and the ferrite sheet after the curing process. This is mainly done due to the fact that larger plates cannot be inserted into the femtosecond laser setup easily. To see if manufacturing conditions would result in a functional degradation, the ferrite sheets are inserted into an oven at 180 °C over four hours with the result that no differences can be detected after this procedure. Therefore, in theory, it is also possible to structure the antenna into the top metal layer after plate manufacturing (Fig. 5.17).

The connection to both, antenna and sensor node, is led out of the edge of the plate, so it is possible to connect different sensor nodes to the same antenna for validation purposes. The antenna design for antennas applied directly onto plates is slightly adjusted to the stack-up mainly investigated in Chap. 3 for concept validation. This has to be done to compensate for the missing layers of handling foil and ferrite adhesive protection that are removed to apply the antenna and ferrite into the cutout, otherwise, the resulting resonance frequency will be too low (see Fig. 5.18). The characteristics of the plate-adjusted antenna geometries are N=5.9 turns, a conductor width of w=2.44 mm, an inner diameter of d_{in}=33 mm, an outer diameter of

Fig. 5.16 Measurement at 50 kHz center excitation frequency inside the first plate with piezoresistive sensor

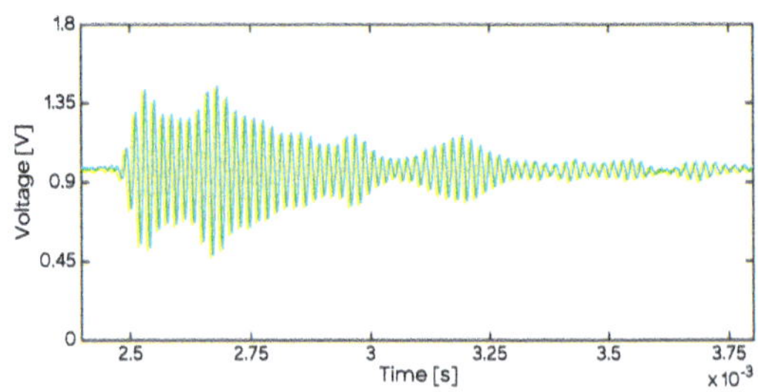

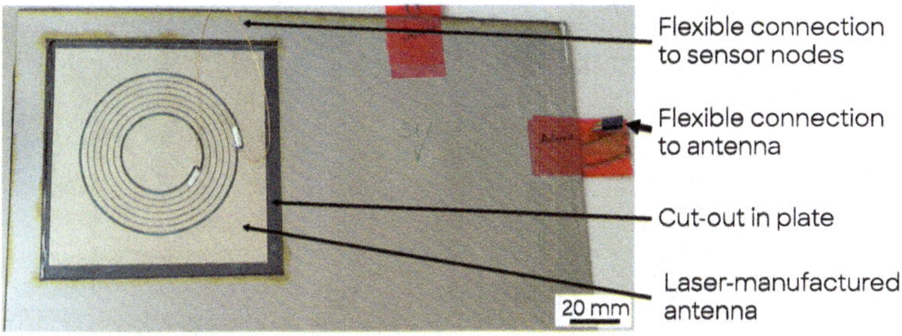

Fig. 5.17 Plate with antenna and ferrite sheet applied after plate manufacturing. Connections toward sensor nodes inside the plate are done outside the plate to allow for flexible system attachment

Fig. 5.18 Measurement of impedance over frequency for different antennas placed in the final application

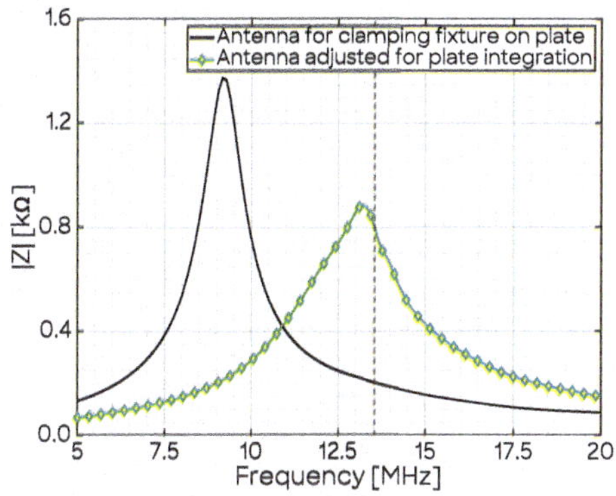

d_{out}=65 mm and thereby a theoretical inductance of L_{theo}=2.21 µH. There are fewer windings and thereby less inductance because the influence of the ferrite sheet is stronger placed this close to the antenna.

The measurement of impedance over frequency shows that with this adjusted design, the target frequency of 13.56 MHz is successfully achieved, although the quality factor is not as high as for the other designs (see Fig. 5.18). Nevertheless, all sensor nodes attached to this antenna show full functionality and readout capability in combination with a reader applied close to the FML-integrated antenna.

5.4.4 CFRP-Stainless Steel Plate, Piezoresistive

After realizing the double amplifier stage for sensor node version 4, a CFRP-steel plate with integrated piezoresistive sensors is used to verify the readout capability with this version now adapted for piezoresistive Lamb wave measurements. The plate is a square-shaped one with an edge length of 500 mm (see Fig. 5.19b). It is manufactured from two stainless-steel layers as top and bottom layer with eight layers of HexPly® 8552 carbon fiber prepreg in between, of which the four layers

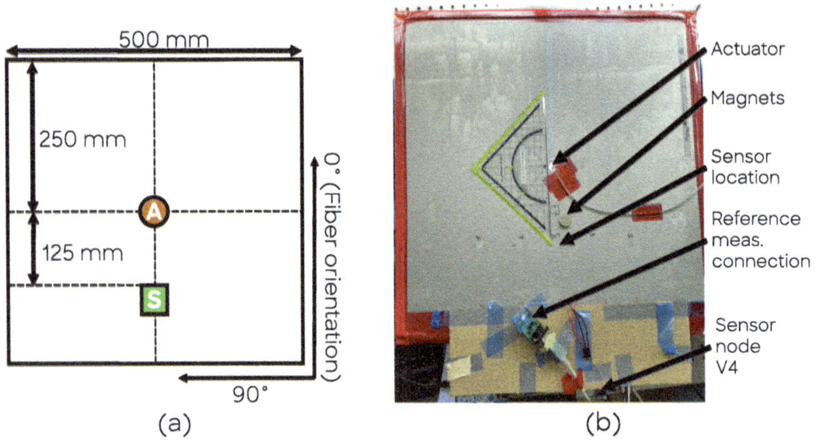

Fig. 5.19 CFRP-steel plate. **a** Schematic measurement setup. **b** Labeled photograph of the plate

in the middle are cut out for sensor integration. Inside the plate, five piezoresistive sensors are integrated from which the middle one is chosen as the active sensor for the presented measurements. It is directly in line of fiber direction with the surface applied actuator (see Fig. 5.19a). As mentioned before, the last version of the sensor node which is used for the presented results is manufactured on a rigid PCB for validation and is therefore not embedded at this point. Nevertheless, a manufacturing on flexible substrate and embedding would follow the same procedure as already presented for the previous node version. For this experiment, the sensor node is connected to the integrated sensors by a wired connection, which at the same time enables the use of a commercial bridge amplifier with the exact same sensor to verify the sensor signal readout with it before connecting the wireless electronics to the sensor. At the same time this provides the opportunity to compare both signals to each other. Again, the software trigger functionality is implemented on the microcontroller to be able to use averaged measurements.

The actuation is done with a 5-cycle hann-windowed sine burst at a center excitation frequency of 75 kHz, to be safely inside the working bandwidth of the sensor. This bandwidth can be found between the first two resonance frequencies of the sensor at 42 kHz (bending mode) and 162 kHz (torsional mode) [26]. As already mentioned, at this point it is possible to compare the measurement performed with a commercially available *DEWETRON DEWE 30–40* bridge amplifier, connected to a high-resolution digital oscilloscope directly to the measurement performed with the wireless sensor node. Both high-resolution, high-resource devices are not suited to be material integrated due to their size, while the low-resource wireless sensor node is intended for material integration. The sensor node measurement in Fig. 5.20 shows the averaged signal calculated from 8 single measurements.

It can be seen that the general signal captured looks quite similar to the high-resource measurement (see Fig. 5.20). The significantly increased noise level is

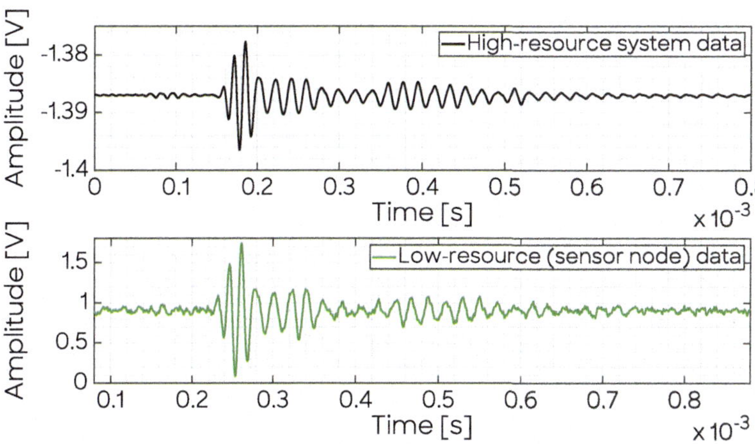

Fig. 5.20 Signal comparison of high-resource measurement equipment to low-resource wireless sensor node

caused by the high amplification necessary because of the low bridge voltage and a noticeable influence of digitization with only 8-bit resolution for the smaller amplitudes. Especially the first part, which following the measurements performed in Sect. 5.4.2 is the most important one to experience amplitude changes due to applied magnets as pseudo defects, can be captured comparably well by the developed low-resource sensor node investigated here. Therefore, in the next step pseudo defect measurements are performed as well for the piezoresistive sensors attached to the wireless sensor node.

Again, magnets are applied and removed as pseudo defects as already done for the GLARE plate before. The pair of magnets is placed 15 mm off the path to the right and in a vertical distance of 85 mm from the actuator (center to center) for this measurement. When placed directly on the path, the signal amplitude is damped too much for the software trigger to work properly. This is due to the relatively high noise level present in the signal due to the high amplification. This can be easily dealt with using an external trigger connection, generated by the AWG-providing the excitation signal, to enable the High Frequency (HF) field and start an exactly timed measurement. This improvement to the system represents the next step and is described in more detail in the outlook section of this work.

The measurement results (see Fig. 5.21) are averaged over eight measurements each. They show similar behavior to the measurements performed with the piezoelectric sensors attached. It can be seen that the measurement signals, although more noisy than in the previous measurements, show a clear response to an applied pseudo damage. If the pair of magnets is applied in proximity to the direct wave propagation path between sensor and actuator, a clear influence can be seen in the measurement. The black line represents the baseline measurement while the green line shows the measurements with applied magnets. The amplitude of the first wave package is significantly smaller than in the baseline measurement and also the following wave

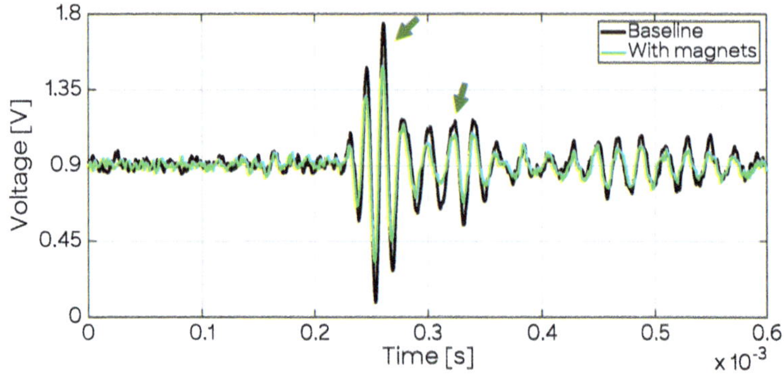

Fig. 5.21 Piezoresistive sensor measurement with sensor node V4

package is slightly damped. The improved version of the sensor node is therefore also able to collect Lamb wave data from piezoresistive Lamb wave sensors from within the material. Although improvements can be made, especially regarding the triggering of the measurement, it is already possible to use piezoresistive Lamb wave sensors with their unique selling point of very small size in combination with the designed wireless sensor node to wirelessly transfer the measurement data out of the laminate.

References

1. H. Towsfyan, A. Biguri, R. Boardman, T. Blumensath, Successes and challenges in non-destructive testing of aircraft composite structures. Chin. J. Aeronaut. **33**(3), 771–791 (2020)
2. D.J. Bull, L. Helfen, I. Sinclair, S.M. Spearing, T. Baumbach, A comparison of multi-scale 3d x-ray tomographic inspection techniques for assessing carbon fibre composite impact damage. Compos. Sci. Technol. **75**, 55–61 (2013)
3. J. García-Martín, J. Gómez-Gil, E. Vázquez-Sánchez, Non-destructive techniques based on eddy current testing. Sensors (Basel, Switzerland) **11**(3), 2525–2565 (2011)
4. M. Castaings, P. Cawley, R. Farlow, G. Hayward3, Single sided inspection of composite materials using air coupled ultrasound. J. Nondestruct. Eval. (17), 37–45 (1998)
5. R. Kažys, A. Demčenko, E. Žukauskas, L. Mažeika, Air-coupled ultrasonic investigation of multi-layered composite materials. Ultrasonics **44**, e819–e822 (2006)
6. W. Essig, Y. Bernhardt, D. Döring, I. Solodov, T. Gautzsch, M. Gaal, D. Hufschläger, R. Sommerhuber, M. Brauns, T. Marhenke, J. Hasener, A. Szewieczek, W. Hillger, Air-coupled ultrasound - emerging ndt method. *ZfP Zeitung - Articles only* (2021), pp. 32–43
7. H. Lamb, On waves in an elastic plate. *Royal Society* (1916)
8. L. Rittmeier, N. Rauter, A. Mikhaylenko, R. Lammering, M. Sinapius, The guided ultrasonic wave oscillation phase relation between the surfaces of plate-like structures of different material settings. Acoustics **5**(1), 136–164 (2023)
9. V. Giurgiutiu, J. Bao, W. Zhao, Piezoelectric wafer active sensor embedded ultrasonics in beams and plates. Exp. Mech. **43**(4), 428–449 (2003)

10. M. Dziendzikowski, A. Kurnyta, K. Dragan, S. Klysz, A. Leski, In situ barely visible impact damage detection and localization for composite structures using surface mounted and embedded pzt transducers: A comparative study. Mech. Syst. Signal Process. **78**, 91–106 (2016)
11. T. Feng, Structural health monitoring of thick composites using ultrasonic guided waves (2022)
12. V.Y. Senyurek, Detection of cuts and impact damage at the aircraft wing slat by using lamb wave method. Measurement **67**, 10–23 (2015)
13. M.S. Hameed, Z. Li, J. Chen, J. Qi, Lamb-wave-based multistage damage detection method using an active pzt sensor network for large structures. Sensors (Basel, Switzerland) **19**(9) (2019)
14. F. Hervin, L. Maio, P. Fromme, Guided wave scattering at a delamination in a quasi-isotropic composite laminate: Experiment and simulation. Compos. Struct. **275**, 114406 (2021)
15. A.A. Ijjeh, S. Ullah, P. Kudela, Full wavefield processing by using FCN for delamination detection. Mech. Syst. Signal Process. **153**, 107537 (2021)
16. A. Migot, H. Mei, V. Giurgiutiu, Numerical and experimental investigation of delaminations in a unidirectional composite plate using ndt and shm techniques. J. Intell. Mater. Syst. Struct. **32**(16), 1781–1799 (2021)
17. H. Pfeiffer, I. Pitropakis, M. Wevers, The use of pseudo-defects in validation of ultrasonic structural-health-monitoring technologies. in *10th European Conference on Non-Destructive Testing, Moscow 2010, June 7-11, NDT.net* (2010)
18. V. Giurgiutiu, *Structural Health Monitoring With Piezoelectric Wafer Active Sensors*. (Academic Press/Elsevier, Amsterdam, 2008)
19. I.A. Viktorov, *Rayleigh and Lamb Waves: Physical Theory and Applications*, vol. A Series of Monographs of *Ultrasonic Technology* (Plenum Press, New York, 1967)
20. J.D. Achenbach, *Wave Propagation in Elastic Solids, North-Holland Series in Applied Mathematics And Mechanics*, vol. 16. (North-Holland Pub. Co, Amsterdam, 1973)
21. T. Salaoru, A. Toader, M. Arghir, M. Andrei, M. Tudose, I. Popescu, C. Donciu, Lamb waves tunning on aluminium plates for structure health monitoring. Incas Bull. **6**(1), 73–81 (2014)
22. S. Vázquez, J. Gosálbez, I. Bosch, A. Carrión, C. Gallardo, J. Payá, Comparative study of coupling techniques in lamb wave testing of metallic and cementitious plates. Sensors (Basel, Switzerland) **19**(19) (2019)
23. V. Giurgiutiu, Lamb wave generation with piezoelectric wafer active sensors for structural health monitoring, in *SPIE's 10th Annual International Symposium on Smart Structures and Materials and 8th Annual International Symposium on NDE for Health Monitoring and Diagnostics* (2002)
24. F.A. Masurkar, N.P. Yelve, Locating a damage in an aluminum plate using lamb waves (2015)
25. U. Patel, Development of measurement setup for damage detection with reproducible lamb waves in frp composites with integrated pwas sensors, *Master's thesis, University of Bremen* (2023)
26. J.N. Haus, W. Lang, T. Roloff, L. Rittmeier, S. Bornemann, M. Sinapius, A. Dietzel, Mems vibrometer for structural health monitoring using guided ultrasonic waves. Sensors **22**(14) (2022)

Chapter 6
Conclusion

A fully digital, self-sufficient, wireless sensor node that can be completely embedded into a highly challenging host material has been presented. It can wirelessly transmit Lamb wave sensor data out of the laminate by using a laminate-structured antenna. The design process of both electronic sensor node and antenna is described in detail covering every design requirement identified at the beginning.

Even though wireless communication using the outer metal layer as an antenna for HF-RFID communication in close proximity to conductive carbon fibers is a challenging approach, it can be implemented by using one additional layer in the material stack made from high-permeability ferrite material. Due to interface transitions of different permeability—air to ferrite and ferrite to air—the magnetic field lines are redirected through this layer and thereby prevented from entering the conductive layer as described in Chap. 3 in detail. As long as the overall design of the antenna is adjusted well enough for the application, a good performance can be achieved for this inter-laminar antenna approach. To achieve a resource-efficient design process for changing antenna requirements, an FEM simulation model is presented with which it is possible to simulate the full material stack and its influence on the antenna performance before manufacturing. The results of the simulation are in good agreement with the measurement results also presented in Chap. 3.

The electronic part of the system is designed in a way that high temperatures and elevated pressure levels during the embedding process do not lead to system failure when manufacturing FML using epoxy-based prepregs. Apart from being resilient to harsh embedding conditions, components are also chosen to be as thin as possible using commercially available packages to keep the influence on the host material as small as possible. Therefore, the sensor node measures 17 mm by 17 mm with a total height of less then 800 μm. The fully digital approach is chosen to be able to process both piezoelectric as well as piezoresistive sensor data, which require different interface- and amplification settings. With this low-resource sensor node,

S. Bornemann, *Development of a Self-Sufficient, Wireless Sensor Node Using the Outer FML Layer as an Antenna*, Mechanics and Adaptronics,
https://doi.org/10.1007/978-3-031-70648-6_6

it is possible to sample data with 8-bit resolution at a maximum sampling rate of 1.23 Msps, fully sufficient for the intended Lamb wave excitation frequency of up to 300 kHz.

Finally, measurements of Lamb waves using the material-integrated wireless sensor node with the two different sensor approaches are presented. The measurements are done within FML plates of different geometries and material configurations and show good reproducibility and sufficient resolution. Pairs of magnets are used as removable pseudo-defects for validation of the resolution and reliability of the embedded system. The influence can be clearly seen in the presented measurements for both sensor types as intended.

Finally, it can be concluded, that the developed self-sufficient, wireless sensor node presented in this work is fully capable to pick up sensor data of Lamb wave measurements directly at the location of event and to transmit it out of the laminate with a sufficient resolution for damage diagnostic purposes. This provides the clear advantage of eliminating the need for connectors and long wired connections to sensors distributed over the whole monitored structure, e.g., an airplane, thereby saving weight, and costs and preventing possible electrical cross-talk problems during measurements from the start.

6.1 Outlook

Within the focus of this work, a sensor node allowing the connection to one single sensor directly connected to it is presented. To enable the system to work as an SHM system over a large area and to be able to locate damage in that area precisely, a sensor network will be necessary. Using the passive RFID approach enables each sensor node to be individually identified among others and therefore to create a sensor network in a different way than it is already used for different applications. During aircraft component manufacturing, it will be possible to keep track of the location of each sensor node which can then be individually recognized by its unique ID. A readout system that exchanges power and data with a sensor node in its proximity can save the sensor response together with the node's ID, enabling damage detection algorithms to read the fixed sensor node location from a database.

An alternative scenario is the direct implementation of damage prediction algorithms on the microcontroller of the embedded sensor node. For this, the system has to be slightly adjusted, i.e., the microcontroller has to be upgraded to a version of higher performance to provide more Random-Access Memory (RAM) for the necessary operations. With this, it would still be a big challenge but a first assessment of appropriate methods presented by Bosse et al. [1] looks quite promising. In addition to the intermediate data storage on the EEPROM, the fast transfer option of the RFID IC could be used in this case for sensor node to enable communication between sensor nodes.

Regarding an improvement of the data acquisition, as already mentioned in Chap. 5, the realization of a hardware-triggered measurement should be investigated. This could be done by developing a combined actuation and readout unit, where the actuation side wirelessly excites the actuator and directly outputs a trigger edge toward the reading part of the system. This side then controls either the activation of the HF-field to always start data sampling at the same time. Alternatively, if the timing performance of this approach is not sufficient, the supplying HF-field could be permanently available while the measurement is started by a command exchanged between reader and sensor node via the already mentioned fast-transfer functionality. With this, reliable averaging would be possible even if amplitudes are low or the signal is too noisy to use the software trigger. Another advantage of this approach is that the time-of-flight data is available and can be used, e.g., as an additional damage diagnostic resource.

Reference

1. S. Bosse, S. Bornemann, B. Lüssem, Virtualization of tiny embedded systems with a robust real-time capable and extensible stack virtual machine rexavm supporting material-integrated intelligent systems and tiny machine learning (2023)

Appendix

A.1 Publications

- **Bornemann, S.**; Haus, J.N.; Sinapius, M.; Lüssem, B.; Dietzel, A.; Lang, W. Stainless-Steel Antenna on Conductive Substrate for an SHM Sensor System with High Power Demand. Sensors 2021, 21, 7841. DOI:10.3390/s21237841
- **Bornemann, S.**; Lang, W. Experimental Study on Stress Impact during FML Manufacturing on the Functional Conformity of an Embeddable SHM-Sensor-Node. Eng. Proc. 2021, 10, 72. DOI:10.3390/ecsa-8-11323
- **Bornemann, S.**; Lang, W. Considerations and Limits of Embedding Sensor Nodes for Structural Health Monitoring into Fiber Metal Laminates. Sensors 2022, 22, 4511. DOI:10.3390/s22124511
- Haus, J.N.; Rittmeier, L.; Roloff, T.; Mikhaylenko, A.; **Bornemann, S.**; Sinapius, M.; Rauter, N.; Lang, W.; Dietzel, A. Micro-Oscillator as Integrable Sensor for Structure-Borne Ultrasound. Eng. Proc. 2021, 10, 81. DOI:10.3390/ecsa-8–11313
- Haus, J.N.; Lang, W.; Roloff, T.; Rittmeier, L.; **Bornemann, S.**; Sinapius, M.; Dietzel, A. MEMS Vibrometer for Structural Health Monitoring Using Guided Ultrasonic Waves. Sensors 2022, 22, 5368. DOI:10.3390/s22145368
- Droste, D.; Krishnappa, L.; **Bornemann, S.**; Ohlendorf, J.-H.; Vargas, M.G.; Herrmann, A.S.; Lang, W. Investigation of the compaction behaviour of a quasi-unidirectional non-crimp fabric during the vacuum infusion process. Journal of Composite Materials. 2022;56(16):2509–2524. DOI:10.1177/00219983221092011
- Bosse, S.; **Bornemann, S.**; Lüssem, B. Virtualization of Tiny Embedded Systems with a robust real-time capable and extensible Stack Virtual Machine REXAVM supporting Material-integrated Intelligent Systems and Tiny Machine Learning. DOI:10.48550/arXiv.2302.09002 (PP)

S. Bornemann, *Development of a Self-Sufficient, Wireless Sensor Node Using the Outer FML Layer as an Antenna*, Mechanics and Adaptronics,
https://doi.org/10.1007/978-3-031-70648-6

A.2 Schematics

Here the schematics of the sensor node design are presented. They are combined to fit on two pages, therefore global ports are still present. The schematic for micro-controller, LDO and RFID IC can be found in Fig. A.1, while the schematic of the amplifier and filter stage is presented in Fig. A.1 (Fig. A.2).

A.3 Software

In the following the main routine embedded C code which runs on the sensor node's microcontroller is presented. It is the version with included software trigger since it also shows the implementation of the simple readout software. The hardware drivers are provided by ST's HAL library and are therefore not introduced here.

```
/* Includes ------------------------------------------------------------------*/
#include "main.h"

#define ADC_BUF_LEN  0x9F6 //Length of ring buffer

ADC_HandleTypeDef hadc; //Define ADC handler label
DMA_HandleTypeDef hdma_adc; //Define DMA channel for ADC handler label
I2C_HandleTypeDef hi2c1; //Define I2C bus handler label

uint8_t adc_buf[ADC_BUF_LEN]; //Ring buffer
uint8_t adc_sum[ADC_BUF_LEN]; //Second buffer for mean value
uint8_t repetitions=1; //Definition for multiple repetitions
uint8_t count_repetitions=0; //Counter for multiple repetitions
uint8_t overflow = 0; //Ring buffer overflow marker
uint16_t m = 0; //Threshold location storage
int16_t n = 0; //Variable for data structuring
uint8_t count = 0; //Counter beginning of signal identification
uint8_t count_in = 0; //Counter variable beginning of signal reset
uint8_t stop_condition_reached = 0; //Stop condition over/under threshold
uint8_t res = 0;  //Intermediate storage of ring buffer value
HAL_StatusTypeDef ret; //I2C status return bit
uint8_t i2c_data[3]; //I2C send data buffer
uint8_t i2c_pwd[19]; //I2C password storage
uint16_t mem_addr[10]; //Memory address array definition

/* Private function prototypes -----------------------------------------------*/
void SystemClock_Config(void);
static void MX_GPIO_Init(void);
static void MX_DMA_Init(void);
static void MX_ADC_Init(void);
static void MX_I2C1_Init(void);
/**
  */
int main(void)
{
```

```
/* MCU Configuration--------------------------------------------------------*/

/* Reset all peripherals, initializes flash interface and Systick. */
HAL_Init();

mem_addr[0]=0x0000; //initialize memory address array
mem_addr[1]=0x00FF;
mem_addr[2]=0x01FE;
mem_addr[3]=0x02FD;
mem_addr[4]=0x03FC;
mem_addr[5]=0x04FB;
mem_addr[6]=0x05FA;
mem_addr[7]=0x06F9;
mem_addr[8]=0x07F8;
mem_addr[9]=0x08F7; //2550 byte

/* Configure system clock */
SystemClock_Config();

/* Initialize all configured peripherals */
MX_GPIO_Init();
MX_DMA_Init();
MX_ADC_Init();
MX_I2C1_Init();

/* Set bridge supply voltage pin to high */
HAL_GPIO_WritePin(GPIOA, GPIO_PIN_6, GPIO_PIN_SET);

/* Verify I2C initialization was successful. If it was not, repeat. */
ret=HAL_I2C_IsDeviceReady(&hi2c1, 0xA7, 100, HAL_MAX_DELAY);
while(ret!=HAL_OK){
 ret=HAL_I2C_IsDeviceReady(&hi2c1, 0xA7, 100, HAL_MAX_DELAY);
}

/* Present Password to RFID-IC */
i2c_pwd[0]=0x09;
i2c_pwd[1]=0x00;
for(int j=2; j<10; j++){
 i2c_pwd[j]=0x00;
}
i2c_pwd[10]=0x09;
for(int k=11; k<19; k++){
 i2c_pwd[k]=0x00;
}
HAL_I2C_Master_Transmit(&hi2c1, 0xAE, i2c_pwd, 19, HAL_MAX_DELAY);

/*Check if password ok: If 0x00 --> I2C security session closed*/
i2c_data[2]=0x00;
while(i2c_data[2]==0x00){
 i2c_data[0]=0x20;
 i2c_data[1]=0x04;
 HAL_I2C_Master_Transmit(&hi2c1, 0xA7, i2c_data, 2, HAL_MAX_DELAY);
 HAL_I2C_Master_Receive(&hi2c1, 0xA7, &i2c_data[2], 1, HAL_MAX_DELAY);
}

/*Write EH settings - only necessary at first DEBUG of new RFID IC*/
```

```
    i2c_data[0]=0x00; //Register 0002: EH_MODE
    i2c_data[1]=0x02;
    i2c_data[2]=0x00; //Set to 0x00 for EH forced after boot
    HAL_I2C_Master_Transmit(&hi2c1, 0xAE, i2c_data, 3, HAL_MAX_DELAY);

    i2c_data[0]=0x20; //Register 2002: EH_CTRL_Dyn
    i2c_data[1]=0x02;
    i2c_data[2]=0x03; //Set to 0x03 to enable EH feature
    HAL_I2C_Master_Transmit(&hi2c1, 0xA6, i2c_data, 3, HAL_MAX_DELAY);
    HAL_Delay(4);
/* Verify I2C transfer was successful. If not, repeat. */
    ret=HAL_I2C_IsDeviceReady(&hi2c1, 0xA7, 100, HAL_MAX_DELAY);
    while(ret!=HAL_OK){
      ret=HAL_I2C_IsDeviceReady(&hi2c1, 0xA7, 100, HAL_MAX_DELAY);
    }

    while (count_repetitions<repetitions)
    {
      /* Start ADC, use DMA, save to ring buffer */
      HAL_ADC_Start_DMA(&hadc, (uint32_t*)adc_buf, ADC_BUF_LEN);
    do
    {
      m=0; //Reset buffer overrun
      do
      {
         res=0;
         res=*(adc_buf + m); //Read value from ADC buffer at location "m"
    /* Check threshold >[0xA9=169(66%)] or <[0x54=84(33%)] */
         if(res > 0xA9 || res < 0x54)
         {
    /* Stop DMA ADC conversion (fills up ~half buffer before stop) */
            HAL_ADC_Stop_DMA(&hadc);
            stop_condition_reached = 1;
         }
         else
         {
            m++;
         }
      /* If no trigger point could be found and end of buffer not reached,
           repeat */
       }while(m<ADC_BUF_LEN && stop_condition_reached!= 1);
      }while(stop_condition_reached!= 1); //If no trigger point could be
           found, repeat

      stop_condition_reached=0; //Reset variable
      count=0; //Set counter variable to 0
      count_in=0; //Set counter for reset variable to 0
      n=m; //To keep value in m, copy to n
      overflow=0; //Set overflow variable to 0

      do
      {
          n--;  //Decrease n by 1
          if(n<0)   //If underflow is detected
          {
             n=ADC_BUF_LEN-1; //Set position to end of buffer
```

```
        }
        if(n==m+1) //Condition when complete buffer has been compared
        {
           overflow=1;
        }

/* Find start of signal, identification criterion that 50 times only
     measurement values within 10% of the reference voltage are present */
      if(overflow==0)
      {
    /* Check threshold between [0x73=115(45%) and 0x8C=140(55%)] */
        if(*(adc_buf + n) >= 0x73 && *(adc_buf + n) <= 0x8C)
        {
           count++; //Increase counter
           count_in=0; //Reset the reset-counter
        }
        else
        {
           count_in++; //Increase reset-counter
    /* Reset counter when three consecutive values differ from the
        "0"-area */
                 if(count_in >= 3)
           {
             count=0;
           }
        }
      }
    /* If buffer has completely compared before start of signal could be
        identified */
      else
      {
         n=m; //Set n again to trigger point location
         count=50; //Set count to 50 to leave loop
      }
     }while(count!=50); //If soon as start has been found (or if no start
         to find) stop loop

     n=n+50; //Jump to location after 50 times low-signal

     if(n > (ADC_BUF_LEN-1)) //Check for overflow
     {
      n=0;    //Reset counter at Overflow
     }
/* Find first true trigger value >=[0xA9=169(66%)] */
     while(*(adc_buf + n) < 0xA9)
     {
      n++;    //If below trigger threshold move further right
      if(n > ADC_BUF_LEN-1) //Check for overflow
      {
         n=0;  //Reset counter at Overflow
      }
     }
/* Add 50 sample points to signal before sensor signal starts
     (pre-trigger t) */
     n=n-50;
```

```
    if(n < 0)   //If underflow detected
    {
     n=n+50;
     n=ADC_BUF_LEN+n-50;  //Beginn n-50 before end of buffer
    }

    if(count_repetitions==0)
    {
       for(int i = 0; i < ADC_BUF_LEN; i++)
       {
          if((n+i) >= ADC_BUF_LEN)
          {
             n=0-i;
          }
          *(adc_sum + i)= *(adc_buf + (n+i)); //(Empty second buffer +
               measurement buffer)
       }
    }
    else
    {
       for(int i = 0; i < ADC_BUF_LEN; i++)
       {
          if((n+i) >= ADC_BUF_LEN) //At overflow
          {
             n=0-i;    //start at the beginning
          }
    /* (Second buffer n + measurement buffer)/2 by shift */
          *(adc_sum + i)= (*(adc_buf + (n+i))) >>1;
       }
    }
    count_repetitions++; //Increase repetitions done count
   }

   /* Write Second buffer via I2C to EEPROM memory of RFID IC */
   for(int i=0; i<10; i++)
   {
     HAL_I2C_Mem_Write(&hi2c1, 0xA6, mem_addr[i], 2,
          (uint8_t*)(adc_sum+(i*255)), 255, HAL_MAX_DELAY);
        ret=HAL_I2C_IsDeviceReady(&hi2c1, 0xA7, 100, HAL_MAX_DELAY);
        while(ret!=HAL_OK){
           ret=HAL_I2C_IsDeviceReady(&hi2c1, 0xA7, 100, HAL_MAX_DELAY);
        }
   }
/* End of measurement, put mmicrocontroller to stop mode */
    HAL_PWR_EnterSTOPMode(PWR_LOWPOWERREGULATOR_ON, PWR_STOPENTRY_WFI);

  /* Infinite loop */
  while (1)
  {
  }
}
```

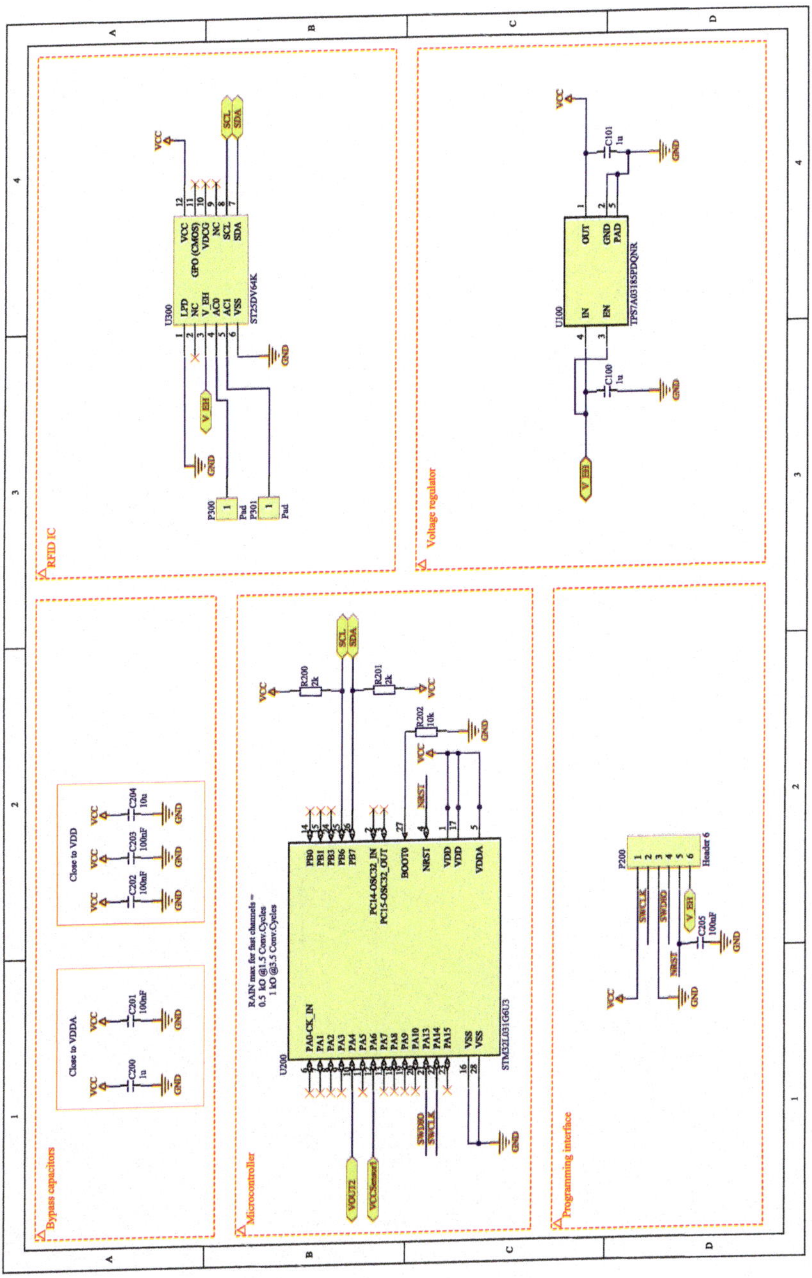

Fig. A.1 Microcontroller, RFID IC and voltage regulator schematic

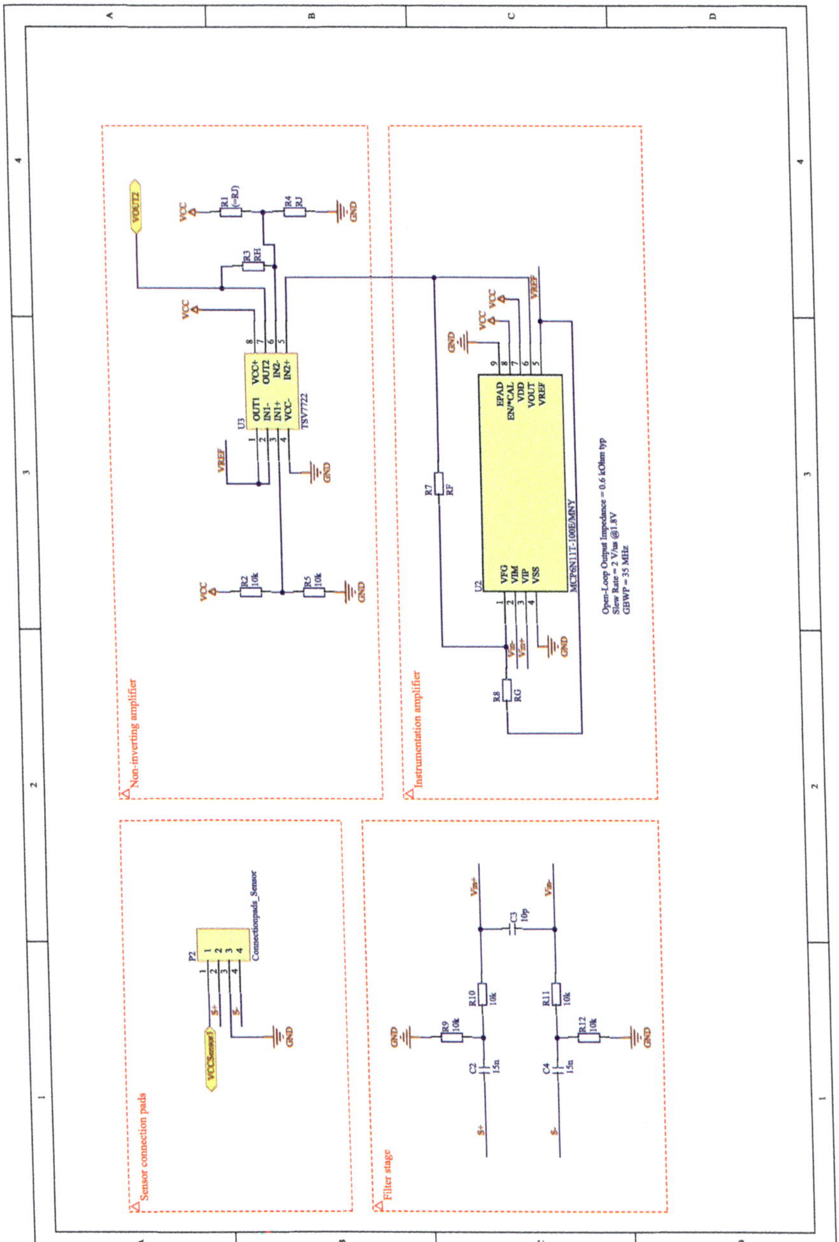

Fig. A.2 Amplifier and filter stage schematic

The manufacturer's authorised representative in the EU is Springer Nature Customer Service Centre GmbH, Europaplatz 3, 69115 Heidelberg, Germany. If you have any concerns regarding our products, please contact ProductSafety@springernature.com

Printed and bound by CPI Group (UK) Ltd, Croydon, CR0 4YY

30/09/2026

02225469-0004